Property
Investment
Theory

T0330830

Property Investment Theory

Edited by

A. R. MacLeary
and
N. Nanthakumaran

CRC Press
Taylor & Francis Group
Boca Raton London New York

CRC Press is an imprint of the
Taylor & Francis Group, an **informa** business

A TAYLOR & FRANCIS BOOK

CRC Press
Taylor & Francis Group
6000 Broken Sound Parkway NW, Suite 300
Boca Raton, FL 33487-2742

First issued in paperback 2019

ISBN-13: 978-0-419-14770-1 (hbk)
ISBN-13: 978-0-367-40055-2 (pbk)

British Library Cataloguing in Publication Data
Property investment theory.
1. Great Britain. Real property
Investment. Analysis
I. MacLeary, A.R. (Alistair R.)
II. Nanthakumaran, M. (Nanda)
332.63′24′0941

ISBN 0–419–14770–5

Visit the Taylor & Francis Web site at
http://www.taylorandfrancis.com

and the CRC Press Web site at
http://www.crcpress.com

CONTENTS

A E Baum, BSC, MPhil, ARICS – Chapter 4

Andrew Baum is Property Research Manager for Prudential Portfolio Mangers Ltd, he is responsible for forecasting property sectors and contributes to the development of property portfolio strategy. He was previously Reader in Property Valuation and Management at the City University and consultant to Richard Ellis. Andrew Baum was founding editor of the Journal of Valuation and a co-author of three books on property investment.

G R Brown, MA, PhD – Chapter 7

Gerald Brown is Director of Property Research Services and Tutor for the City University MBA Programme; he holds a PhD in Finance, a Masters degree in Business Studies and a degree in Architecture and has wide experience in the property industry, being involved in a number of major development projects. More recently, he was responsible for developing investment research ideas for Richard Ellis. His main interest lies in the performance measurement of property portfolios and he subsequently established Property Research Services to develop this area. Dr Brown is an external examiner for the MSC Project Management course at the Polytechnic of the South Bank.

Professor W D Fraser, BSc, MSc, PhD, FRICS – Chapter 2

Will Fraser is head of the Department of Land Economics at Paisley College. He was formerly a lecturer at the City University, London and previously a partner in Bingham Hughes and Macpherson, Chartered Surveyors, Inverness. He is the author of Principles of Property Investment and Pricing.

J T Hetherington, BSc(Hons), FRSS – Chapter 6

John Hetherington is Chief Research Analyst at Hillier Parker. A graduate of Liverpool University where he obtained a degree in Computational and Statistical Science. He has a particlar interest in mathematical model building and how this can be applied to portfolio analysis, forecasting property values and the identification of areas of investment. He is responsible for a number of papers on portfolio analysis as well as the production of publications such as the ICHP Rent Index and Average Yields. He represents Hillier Parker on the technical committee of the Property Index.

Professor A R MacLeary, MSc, DipTP, FRICS, FRTPI, FRSA – Chapters 1 and 9

Alistair MacLeary is the MacRobert Professor of Land Economy at the University of Aberdeen and recently served as the Dean of the Faculty of Law. His very extensive experience spans both the public and private sectors in the planning and development field and includes

work on the Sullom Voe Master Plan, the UN Suez Canal Regional
Structure Plan and as a UN Adviser for the Guatemala Earthquake
Disaster. His research activity has included a two-year programme
examining regional shopping centres with a view to formulating
planning policies for the Department of the Environment. Professor
MacLeary serves on various RICS, RTPI and CASLE committees. He is a
past Chairman of the Planning and Development Division and a past
Chairman of the Watt Committee on Energy. He has served on the
Northfield Committee and on one of the statutory committees of the
Forestry Commission. He is editor of <u>Land Development Studies</u>.

S J E Morely, MA, BSc, ARICS, DipTP - Chapter 5

Stuart Morely is Head of the School of Estate Management at the
Polytechnic of Central London with teaching responsibilities for
property investment and development, valuation and appraisal. He
undertakes consultancy work for both the private and public sector
and contributes to various books on development and investment
appraisal, office development and the future of the property market.

N Nanthakumaran, BSc, ASVA - Chapters 3 and 9

Nanda Nanthakumaran is a lecturer in Land Economy at the University
of Aberdeen specialising in valuation and quantitative studies.
After obtaining a degree in Estate Management from the University of
Reading he worked in Sri Lanka and Nigeria and was a visiting
lecturer in Valuation at the Universities of Sri Lanka and the
Copperbelt, Zambia.

Professor C W R Ward, MA, MA, PhD, AMSIA - Chapter 8

Charles Ward is Professor of Accountancy and Deputy Principal at the
University of Stirling. After reading Economics at Cambridge
University he worked in industry for six years before entering
teaching. Whilst at Leicester Polytechnic he specialised in Land
Economics and the property market. This led to a postgraduate degree
in Finance at Exeter University and a development of interest in
invstment markets. In 1979 he completd a PhD in Property Investment
at Reading University whilst lecturing at Lancaster University.

ACKNOWLEDGEMENTS

The success of any seminar depends not only upon the quality and content of papers and their presentation but also on the quality of discussion, comment and criticism from the delegates. The seminar was attended by delegates who were invited by the sponsors. The absolute number attending and the degree of representation of various interest groups obtained was gratifying to the organisers and resulted in highly informed debate. The editors are therefore very grateful to these participants for their most significant contribution which justifed both the holding of the seminar itself and the work of the authors. The names of those attending the seminar are listed in the Appendix.

Clearly a particular debt of gratitude is due to the authors of the papers delivered and discussed at the seminar. Each author interpreted his brief well and the content of each paper lived up to and beyond the expectation of the editors. The responses of delegates at and following the seminar confirmed the degree of satisfaction which the authors' works produced. The sum of the parts also managed to engender a greater whole and this could not have happened without the active interpretation by the authors of the purposes of the whole seminar as well as their own contributions. This volume represents a small reward for their efforts.

Thanks are most certainly due to those delegates who happily took on the role of reporters and who were therefore instrumental in focusing discussion on the principal issues identified in the latter part of the seminar. This process of distillation made the editors' task of drawing together the collective thoughts of presenters and delegates in the final chapter much easier. The reporters were Dr N Crosby, Department of Planning and Land Management, University of Reading, A McIntosh, Healey and Baker, Chartered Surveyors, London, D Gillespie, Cluttons, Chartered Surveyors, London, G Morrell, Prudential Portfolio Managers, T Dixon, College of Estate Management, Reading, Dr S Hargitay, Department of Surveying, Bristol Poytechnic and D Carr, Hillier Parker Financial Services.

The editors are grateful to the Partners of Hillier Parker for their support for the seminar and in particular to Professor Russell Schiller, an Honorary Professor in the Department of Land Economy at Aberdeen, who was critically involved in the inception and execution of the seminar.

Sue Nickson of Hillier Parker was responsible for the detailed organisation. Our thanks to her for undertaking this considerable burden. Our thanks also to Maureen Reid for her painstaking work in preparing the typescript.

Seven of the nine chapters of this volume were presented at a two-day
seminar entitled 'Advances in Property Investment Theory' which was
held in London in February 1988. The seminar was organised by the
Department of Land Economy at the University of Aberdeen and the
event was supported by Hillier Parker, Chartered Surveyors of London.
     The reason for holding the seminar was because of the increasing
concern which was being expressed within academic communities and by
practitioners and researchers in private practice about the adequacy
of those techniques which are being used in the valuation and
appraisal of property investments. Allied to this concern was the
uncertainty apparently attaching to the appropriateness of capital
market theory to the analysis of property investments.   Some writers
were arguing that property should adapt new and more appropriate
theories and practices in capital market theory.
     In order to address these issues properly it was felt necessary
to hold a seminar which would provide the opportunity to bring
together all of those concerned with developing and applying the
methodology used in investment appraisal and property valuation. The
invited group was carefully drawn from leading practitioners and
academics representing the most significant institutions and firms
involved directly in property investment, in the giving of advice on
property investment and in conducting research in this field. The
list of delegates attending the seminar is included in the Appendix.
     To do justice to such an important topic the seminar was held
over two days. The seminar papers were circulated to delegates
before the event in order to allow the maximum opportunity for
reflection of the issues and to encourage the bringing forward of
points for discussion.
     For similar reasons the authors of individual papers were asked
to respond to a brief prepared by the editors and each author's
response to the brief was circulated to the others prior to the
drafting of papers so that maximum compatibility between the papers
could be achieved with the minimum of unnecessary overlap. The
authors responded very well to this direction and their collective
efforts were therefore highly successful in focusing the various
aspects of topics for debate.
     These approaches to the seminar had the effect of both
stimulating and focusing the debate and this made for very active and
clear discussion of the principal issues which were themselves
quickly identified.
     This volume therefore attempts to encapsulate both the essential
messages from the individual papers given at the seminar together
with the essence of the discussion arising out of debate on the
papers.
     In Chapter 1 Professor MacLeary introduces the general background
against which the debate takes place. Essentially this is a brief
appraisal of the developing role of property as a medium of
investment (particularly for large financial institutions) in the
United Kingdom. Resulting from the scale and degree of interest in
property as an investment there is an increasing expectation from the
investment community that property will become more amenable to

analysis and to comparison with other investment media. This demand
serves to highlight the nature of the problems which are addressed in
the remainder of the text.

In Chapter 2 Professor Fraser addresses the ongoing debate about
the efficacy of property market valuations vis à vis investment
appraisal. In this chapter he draws a distinction between market
valuation and assessing value to the individual investor. After a
refreshingly clear appraisal of the methods and their purpose he
concludes that the optimal method of any particular case depends on
the purpose of the valuation and the market evidence available.
However, he is careful to go on to make it clear that the choices of
method are not mutually exclusive and recommends that any disparity
between results should be carefully investigated in order for the
best judgemental decision to be made.

In Chapter 3 Mr Nanthakumaran examines some of the problems
involved in the use of contemporary discounted cash flow models for
property investment and analysis. Their applications, the
assumptions on which they are based and their limitations are all
carefully and critically analysed. He concludes that while current
models may be adequate for the valuation of standard freehold
reversions they are not adequate in their present form for the
valuation of distant reversions or geared short leases. The critical
deficiencies lie, firstly in the use of the same discount rate for
the rack rented freehold and for the reversion and, secondly, in the
use of a long term implied growth rate in valuing geared short
leaseholds.

Chapters 2 and 3 may fairly be said to be representative of the
most up to date thinking in the field of property valuation. Indeed
they demonstrate that the need for research must now move into the
areas of risk analysis and forecasting in order to further the state
of the art in property investment valuation.

In Chapter 4, Mr Baum undertakes an in-depth analysis of the
problems relating to the measurement of the impact of depreciation on
commercial property. After defining the nature of depreciation and
the particular and diverse ways in which it makes an impact on the
financial returns from property he then deals with the theoretical
and empirical approach to the means of measurement and estimation of
depreciation, giving in the process a rigorous review of the research
studies which have been undertaken into the subject. In conclusion
he argues that the incidence of depreciation in property is a
significant variable which underpins further the perceived need for
an explicit approach to performance measurement and investment
appraisal.

In Chapter 5, Mr Morley approaches the problems inherent in the
analysis of risk in the appraisal of property investments. He
commences by highlighting the need for risk analysis in property
investment and proceeds to detail those variables prone to risk in
property investment and the relevance of risk analysis to different
types of property investment. He then considers the way in which the
property market currently deals with risk bfore going on to detail
the specific techniques available for the analysis of risk together
with an assessment of their respective problems. While recognising
that the cruder the method, the easier it is to use in practice and

the more sophisticated the method the more difficult it is to use in practice he concludes that simulation techniques generally (and Monte Carlo simulation in particular) are justified in themselves and not least because they lead to a specific exposure of those variables subject to change in the decision making process.

In Chapter 6, Mr Hetherington discusses the need for forecasting and the techniques available for this purpose. Like other authors he points initially to the implicit assumption made within 'all risks' yields used for valuation purposes and confirms therefore that forecasting is not a new phenomenon in property valuation; the innovation, he reminds us, is in the application of mathematic models for the purpose. He then discusses those variables relevant to the measurement of property performance (yield, rent, rental value, capital value and rate of return), and goes on to consider the generation of forecasts for yields and rents in more detail. By demonstrating the way in which demand from occupiers for property is a key variable in forecasting rentals and demand from investors for returns is a key variable in forecasting yields he is able to illustrate the way in which such variables may be measured using the tools of economic and statistical analysis. The use of such techniques can smooth trend lines and identify·turning points and, subject always to the difficulties created by the paucity of information which may be available, they can be an invaluable instrument of property investment analysis.

In Chapter 7, Dr Brown discusses the development and application of portfolio theory to property investment analysis. He introduces the subject by examining the development of capital asset pricing theory – with its emphasis on the measurement of market risk. Developing from this base he shows that rational investors will wish to maximise net present value and that the correct valuation of property assets will be present value. He then addresses the problem of asset selection in relation to the notion of maximising net present value and he discusses the limits of diversification obtainable from a property portfolio, with reference to some empirical evidence.

In Chapter 8 Professor Ward examines the use of those asset pricing models developed within theories of finance and discusses their application to property investments. Property investments are considered primarily as long term investments and the attraction of property to investors is analysed using that perspective. He examines the results of research into bonds and equities (as long term assets) and concludes that the findings have direct relevance for property investment analysis.

Chapters 7 and 8 form a most useful purpose coming as they do after those chapters which are primarily concerned with the means of handling the analysis of those variables distinctive to property investments. They serve to bring closer the concepts which, until now, have been typically discussed separately by those concerned with property management and valuation and those concerned with the wider analysis of financial assets. Indeed it could be argued that they close the gap between property investment analysis and the analysis of other assets. As Professor Fraser points out in Chapter 2 there is entire consistency between models for the market valuation of property investment and models for investment appraisal – 'Both the concepts and models are flawless'. In the same way the authors of Chapters 7 and 8 present a convincing case for the uniform application of already developed financial theory to all investment assets, including property.

Chapter 9 is an attempt by Professor MacLeary and Mr Nanthakumaran to synthesise the issues arising out of the debate ensuing from the delivery of the papers at the seminar. In doing so they draw heavily upon the work of the reporters who helped to distill much of the discussion. (The names of the reporters are listed in Acknowledgements.) Inevitably, nevertheless, a degree of selectivity and possibly even partiality might have crept in. However, there can be no doubt that the issues examined can stand as those requiring further research. Equally the areas identified for further research may not be all inclusive but almost certainly do represent those with a fair degree of priority and of practical possibility.

Clearly the editors accept full responsibility for any omissions or inaccuracies resulting from their interpretation of the work of the authors and reporters and the contributions made by the delegates at the seminar.

CHAPTER 1

PROPERTY AND INVESTMENT

1   Property and Investment in the UK

The property market in the United Kingdom is a vigorous one and it is
one which has undergone distinctive patterns of change. The dynamics
of change continue and apparently accelerate. The volume of funds
invested in property have grown by an enormous amount over the last
twenty years. Much new investment over the last decade has been into
residential mortgages and this reflects the quickening pace of owner-
occupation encouraged by the present Government. Investment in
commercial property (with which we will be concerned) has shown less
spectacular but nevertheless impressive growth.

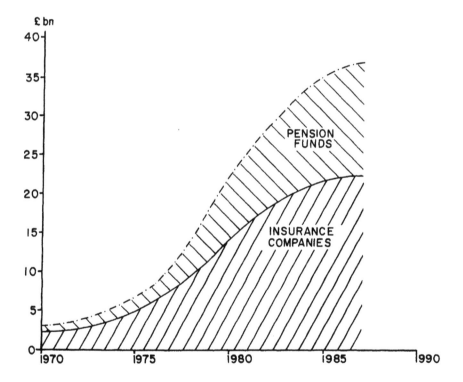

Figure 1   Institutional Investment in Property

Together with equities and government securities property has establ..shed itself as a major component of institutional portfolios subject to continuous scrutiny of its comparative advantage against these other investment media. As a consequence property is now a subject of considerable importance to financial institutions, as well as to property companies and banks.

However, in achieving this status with the investment community property has ridden a roller-coaster courtship of heady euphoria followed by deepest despair. It has changed suitors on a number of occasions according to their perception of its dowry or its perception of their virility. It remains enigmatic in respect of many of its characteristics and stubborn in its refusal to submit to detailed analysis.

The reluctance of property to reveal explicit facets for precise determination has not always been the fault of property. Investors have often enough been prepared to seek a hasty marriage of convenience without much regard to the marriage contract. But property is now firmly wedded into the family of investment media. Maturation and respectability permit the giving up of secrets and a greater willingness to accept accountability. Indeed it may be suggested that if property does not satisfy investors with the information they require then it risks being spurned in favour of competing investments which may look more attractive to investors, particularly in the short term.

In making this somewhat frivolous analogy with the flighty temptress who inevitably becomes the respectable lady it is possible nevertheless to highlight the nature of the property investment dilemma and to isolate those characteristics of property which have given and often continue to give difficulty or to be misunderstood.

In order to set the scene for the particular issues addressed by later chapters it will be necessary to outline briefly the history of investment in property and to consider the nature and behaviour of the principal investors in property.

2 The History of Property Investment in the UK

A number of writers have provided clear and concise accounts of those significant events which have happened to or have been caused by property in the recent past. (Darlow and Brett: 1983, Fraser: 1984, McIntosh and Sykes: 1985). There is general agreement amongst these writers that the main events included the following:

1965-73: the property boom

A distorted property market (essentially caused by the Control of Office and Industrial Development Act 1965) resulted in under supply of commercial space and spiralling rental values. As a consequence of this overheating of the market property companies became subject to take-over bids (encouraged also by the Capital Gains Tax Act 1965 and the introduction of Corporation Tax). The movement of pension funds into property (see p.4) encouraged the boom which was further fuelled by the expansionist policies of the then Chancellor of the Exchequer, The Hon. Anthony Barber. His budget of August 1971 unleashed credit expansion and inflation. British industry failed to respond to the opportunity for cheap capital investment and the banks lent instead to property companies. Such lending almost quadrupled over the four year period (£800m start 1972 to £3000m end 1975) and the rapid growth in lending to property companies emanated in particular from secondary banks. In retrospect this sharp upward trend in debt financing was the main destabilising factor in the commercial property market. But initially high rates of inflation (moving from 8% in 1971 to 20% in 1973) with consequential rises in rack rents together with high rentals at the margins of the office market on uncontrolled prime space encouraged a belief that (already overvalued) properties would sustain capital values at a rate of increase higher than inflation. Rates of interest were not seen as being critical in these circumstances and the secondary banks became increasingly committed to property and hence lending long while borrowing short. But that did not matter - property was a good investment!

1973-74: the crash

In late 1973 the established economic order was thrown into confusion by quadrupling oil prices thus further increasing inflation and critically undermining economic policies designed to enable a 'dash for growth'. Within days minimum lending rates shot up to 13%, severe credit restrictions were brought in and development gains tax was introduced. As a direct result of these economic shocks the consumer boom came to an abrupt halt and with it the demand for commercial floor space. Property companies were now fearfully exposed. They were faced with steeply rising short-term interest rates while their rental income remained static. They could not obtain further borrowing and they could not sell. Companies caught with large development programmes were particularly hard hit because, in the midst of all this adversity, they were faced with rapidly escalating construction costs. Capital values fell sharply. The property companies were highly geared. It was essential that they liquidate assets to repay debt. But there were no buyers. Financial institutions who might have bought cheaply were getting a better return on the money market (with high rates) and probably took the view that, in the circumstances, property would become even more cheap. Unable to sell and raise cash property companies defaulted on loans. Secondary banks unable to recover debt became bankrupt. On the 30th November the London Stock Exchange suspended trading in the shares of London and County Securities. The Bank of England was constrained to undertake a rescue operation to save the secondary banks - and hence the credibility of the City itself.

The story may not be an edifying one from any particular viewpoint but one can appreciate how it comes about that property is seen as the bête noir in the process. It is certainly the case that ignorance of those factors which affect property investment and/or an inability to measure such effects was a major contributor to many of the unwise investment decisions that were made.

<u>1974-82</u>: the expansion of institutional investment

By far the most dominant investors in property are the financial institutions. Insurance comanies and pension funds historically invested in fixed interest securities (because of the restraints of Trust Deed rules). However, since the last war institutions increasingly invested in property as it became clear in the 1960's that growth in both income and capital could produce a greater overall return than fixed interest securities – hence the emergence of the 'negative yield gap' where apparently riskier equities were displaying lower yields than gilts but where the redemption yield would be higher.

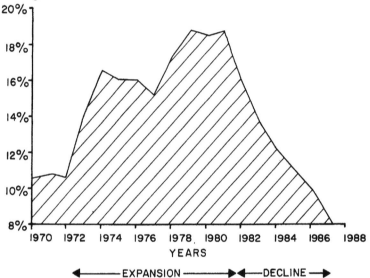

Figure 2   Insurance companies' and pension funds' property
            investment as a percentage of net assets

For the same reasons financial institutions began to show an increasing interest in property as they saw both rents and capital values rise, particularly during periods of high inflation. There were other reasons for increasing their involvement in property. The sheer volume of money flowing into these institutions meant that any investment medium was in demand (e.g. the amount of savings handled by the financial institutions in 1957 were £7bn, by 1987 it was £36bn). During the sixties the practice of commercial property development moved towards the introduction of shorter occupation leases and more frequent rent reviews. The imposition of credit restrictions caused property companies to turn to the financial institutions. Arrangements such as sale and leasebacks were proving to be attractive to both parties. The cumulative effect of such beneficial changes in the property market was a greatly increased interest in investment in property.

However, until the 1970's the financial institutions, interest in property was mainly expressed by the acquisition of shares in

4

property companies. After the crash of '74 it was possible for
institutions to increase their property holdings by purchasing at low
cost the assets of those property companies which had got into
difficulty. Direct investment in property was seen to carry with it
the benefits of greater control of the asset and greater security of
an income which was untaxed in the hands of the pension funds and
life insurance funds.

These factors culminated in a period of significant acquisition
of property investments by financial institutions in the period
1975-1982. During that period the total property assets of financial
institutions rose from £7bn to £27bn. The proportion of property
investment as a part of the total portfolio of financial investments
was at or near 20% during this period.

1983-87: the maturing of the property investment market

Since 1982 quite significant changes have occured in the property
investment market. The total value of the property assets held has
continued to increase over this period but the rate of increase has
been very much less rapid. During this period net annual investment
in property has fallen from about £2bn per annum to £1bn per annum
and the proportion of institutional portfolio structure has also
roughly halved over that period (from 20% to 10%), with the share of
net annual investment falling to 5%.

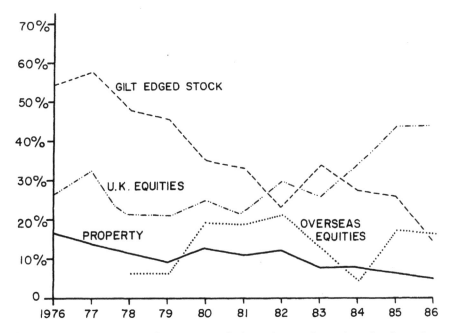

Figure 3   Institutional net annual investment (pension funds and
            insurance companies)

A number of reasons for this apparent disenchantment with commercial property investment can be readily identified. Some funds may have already achieved their target for long-term property assets in their portfolio. During this period there has been sustained a relatively low rate of inflation combined with an unprecedentedly prolonged bull market in equities shifting comparative advantage from property and gilts to equities. This has undoubtedly been the principal reason, that is, the low return available from property relative to other investments. As well as the boom market in equities in the United Kingdom, overseas investments were running at a high level. With high rates of interest and real rates of return short term assets have been more attractive.

Indeed during 1987 pension funds became net disinvestors in property for the first time although the commercial property market has remained buoyant. This relative disenchantment with direct property investment by financial institutions also introduced factors other than concern about rates of return and opportunity cost. Doubts were emerging about the suitability of property as an investment. These doubts focused on several areas.

Firstly, there was concern about the method of valuation of property assets. Leading analysts were concerned that traditional methods of valuation may have proved to be not as useful as more explicit discounted cash flow analysis in determining worth at any given time. Certainly the crash of '74 and its repercussions caused considerable disquiet in this regard (see Greenwell Montague: 1977, for example) and the debate continues to the present time (see Chapters 2 and 3).

Secondly, there was increasing disquiet about the illiquidity of property investments. This referred in part to the cutting back of investment where performance is poor (such as agricultural land in the recent past), but also to the fact that, even for large financial institutions, direct property comes in large indivisible amounts.

Thirdly, there was increasing disquiet about the incidence of depreciation and/or obsolescence in property and its financial implications for the investment. An original article by Bowie:1982, claiming a reduction in yield from a prime office investment from 4.5% to 3.9% net of depreciation, is reputed to have had a direct effect on property investment activity. Subsequent work by Debenham Tewson and Chinnocks:1985, Salway:1986 and Jones Lang Wootton:1987a has further clarified the nature of obsolescence and its measurement. In Chapter 5, Baum uses evidence from a City University study to confirm the nature and characteristics of depreciation and to suggest that identification of depreciation as a third variable is a further argument in support of explicit appraisal of property investments.

During this period financial institutions were therefore becoming more immediately aware of the unique characteristics of property. They had always understood that imperfections in the property market included lengthy and comparatively expensive transfer times and costs, or that property markets were local and often thin and volatile. But they were now becoming increasingly anxious that details of transactions in the market were not easy to come by or that there was no readily available index of current or past prices.

As a consequence it was difficult to analyse risk in property investments.

To be sure the leading firms of chartered surveyors had begun to produce indices of rental, capital and yield performance to attempt, however inadequately, to fill the gap which the absence of an 'All Property Index' designed to supplement other accepted indices has left, and which could never really be filled because of the very nature of property markets.[1] While these indices are useful and while they are primarily concerned with institutional investment, they display different approaches from data bases of different size and quality. As has been pointed out by analysts such as Rowe and Pitman:1982, these differences necessarily have to be kept in mind when making comparisons between any of them and a funds own portfolio of property investments. Efforts by the Investment Property Data Bank:1986 to pool the resources of all of these (and other) firms have met with limited success to date.

Financial institutions as the dominant investors in property found themselves in an uncomfortable position so far as the analysis of risk in property investments is concerned. The appropriate techniques and their application to property investments appeared to be open to question. If an explicit analysis was to be made then particular variables, such as growth in rental income, would require to be measured and there was continuing lack of agreement as to just how this should be done. Techniques requiring determination of statistical outcomes appeared to be even more remote in terms of their application (see Chapter 5).

In the meantime, analysis of property performance was hampered by the paucity of forecasting of essential variables such as rents and capital values and in any case performance measurement would necessarily be handicapped in the continuing absence of an 'All Property Index' (see Chapter 6).

The cumulative effect of such apparent inhibitions was to throw doubt upon the suitability at all of applying conventional means of investment analysis and measurement to property investments. Would it not be more appropriate to approach the investment appraisal of property from a different viewpoint; one that permitted the idiosyncrasies of the property market? While not necessarily being non-substitutable, conventional wisdom suggests that there is insufficient justification for manufacturing (artificially?) different tools of analysis for property investment – rather more careful interpretation should be made of the results of those conventional tools which have a sound theoretical base and which have a degree of credibility in the analysis of other assets (see Chapters 6 and 7).

---

1. Typical of such indices are the Jones Lang Wootton Property Index (capital and rental) the Richard Ellis Property Market Indicators, the Michael Laurie/EIU Property Performance and Average Rents/Indices and the Investors Chronicle/Hillier Parker Rent Index.

1987 and beyond: the crash of '87 and the future

At this point in our brief historical analysis of property investment and the cursory examination of the problems inherent in the process we risk leaving an impression that investors in property were disillusioned and that therefore property markets were in retreat.

During the period 1983 to 1987 institutional investors were not deterred by the peculiarities of the property market and problems of performance measurement, rather they were encouraged by the comparative advantages of the other investment opportunities. As has been mentioned above, property markets remained virile. Throughout this period there was a rapid expansion of commercial property markets to meet healthy consumer demand. The rate of new building to satisfy rising consumer demand was high. Because of the shortage of finance from the financial institutions development companies have been turning to the corporate sector and to the bankers. Low projected rates of inflation – a factor which had discouraged

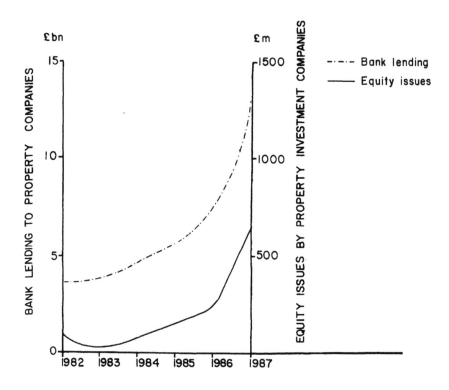

Figure 4   Bank lending to property companies and equity issues by property investment companies.

8

financial institutions from investing in property - made it easier to issue capital debt. Nevertheless the most important single source to fill the supply gap left by the financial institutions has been the banking sector. In deregulated financial markets lenders were finding their margins squeezed because of increased domestic and overseas competition in the corporate sector. Seeking returns comensurate with the risks perceived to be attached to property, the banks have been glad to oblige the property companies.

While there has been a mutuality of interest and while it is salutory for the property industry to remember that there are a variety of potential investors with different motives and expectations in differing sets of circumstances, the apparently astronomical rise in bank lending to property companies, coupled with unsustainably high levels of rental growth in commercial property, inevitably invited comparison with the crash of '74.

There has been concern that the same set of conditions was recurring - with over exposure by lenders to property comanies. Fortunately the comparison is superficial. Although bank lending to property companies is now at an all time high the banks are not vulnerable secondary banks but major UK and overseas banks. The borrowers are not highly geared property companies (including new adventurers) but substantial and established property companies who have survived previous storms and with debt ratios well in control (total borrowings rose from 23.9% of total assets to 27.3% between 1982-87). Investment decisions are no longer made on unrealistic (and unrealisable) expectations of consumer demand and commercial rental values. The actual financial arrangements entered into between the parties are rigorously tested.

The underlying reasons for the crash of '87 are in any event fundamentally different from those which precipitated the crash of '74. However, the interesting aspect of this latest financial hiatus so far as this volume is concerned is the manner in which property now performs as part of an investment portfolio and how its performance may now be analysed as the property sector is increasingly drawn into the mainstream of financial markets.

Before doing so, however, it will be useful to consider what effects the collapse of the Stock Market in 1987 may have had on property yields. On this occasion the property market appears to be enjoying a boom. Rental growth is strong (and is predicted to remain so). The outlook for equities has dulled and there has been a distinct shift in investor interest from equities and into gilts and property. As a consequence equity yields have risen sharply while property yields have remained steady and indeed may be under pressure to move downwards in the future. Whereas, therefore, there was a fall of 27.2% in the FTA All-Share Index over the last quarter of 1987 and an annual return of only 7.9% property investments produced an annualised return of 18.3% (the highest for six years) with industrial capital values and retail rentals showing the highest growth (Jones Lang Wootton:1987b, Warburg Securities:1988, Healey and Baker:1988). Indeed at the end of 1987 those financial institutions which were more heavily committed to property in their portfolios were popularly reported to be advertising the fact with some satisfaction.

3 The Property Market and Property Investment Theory

The purpose of the foregoing historical review has been to put into context the more detailed and specific content of the chapters which follow.

Pre-1974 property was seen to be a useful diversification and risk spreader for rapidly growing investment funds when inflation was increasing and when property was seen to be making its own contribution to target returns in comparison with other assets.

The crash of '74 afforded further opportunities for expansion in property investment until the comparative advantage of equities took over in 1983. In the meantime, certain of the disadvantages, or inhibitions, which property presented to investors were becoming increasingly exposed. These included problems related to methods of asset pricing, to the measurement of variables necessary to satisfy explicit asset pricing models such as risk premium, income growth, liquidity premium and obsolescence. These difficulties themselves call into question the efficacy of models of prediction for these variables. This in turn invites examination of the question as to how the analysis of property investments fits into general theories of finance.

After the crash of '87 there could no longer be any doubt about the robustness of commercial property as an enduring element of the portfolios of large scale investors. At the same time the running debates which had been separately pursued on various factors of property investment analysis had been running out of steam or reaching impasses. In any event it was now clear that property investment had to be more clearly justified to investors than had been the case in the past. The property sector also must be capable of useful comparison with other assets and opportunities for investment. The measurement of variables affecting yield analysis, the method of analysis itself and the theoretical and practical appraisal of individual properties and portfolios had finally to be submitted to tests of acceptability and applicability.

Nor is it simply a question of the present being an appropriate time to address these questions. It is also quite necessary for thinking and practice in property investment to be made explicit. If the profession of the land fails to meet this challenge it will have lost any authority to exercise leadership and enjoy commercial benefit in the practice of property investment. While the Royal Institution of Chartered Surveyors has been showing distinct signs of willingness to orientate the principal landed profession in the country towards broader management processes and the wider finanical community, practitioners themselves show a distinct reluctance to embrace the theory and technology which will lead them in that direction (see Management Analysis Center: 1985). It has been observed that since the advent of electronic calculators having logarithmic and exponential functions the need for tables providing financial factor multipliers has been wholly redundant. Yet the profession itself and, more worryingly, very nearly all undergruadate courses leading to exemption from the professional examinations conspire to continue dependence on valuation tables and on unnecessarily quaint mathematical notation and customary practice oblivious to the uniform application of similar financial mathematics in related fields.

Defence of traditional methods on the grounds that these are customary practices necessary for the market to function through a common understanding becomes increasingly hollow over time. They do not do credit to a learned profession.

The problem of analysing property investments therefore includes an attitudinal reluctance by the practitioners involved and this clearly compounds the not inconsiderable technical difficulties which already obtain. Significiant although this factor must be it is not subject to futher analysis in this book. It is a matter to which the profession itself must look.

Rather we are concerned here with those theoretical and empirical matters which must be addressed in order to help rationalise property investment. The remainder of the book therefore addresses the following topics.

Firstly, it seeks to compare conventional methods of property valuation with more explicit models in order to advance a rationality for the use and purpose of these methods, and to highlight any perceptive or practical difficulties in their implementation. Clarity in determination of asset price is a priority if new debates on this area of uncertainty are to be avoided in the future. (See, for example, Hager and Lord: 1985 and Sykes:1983.)

Secondly, it deals with the identification of those variables inherent in an explicit analysis of property investment and examines the way in which these variables may be quantified by forecasting methods and the way in which the risk of differing outcomes may be addressed by risk analysis.

Thirdly, it addresses the relationship between property as an investment medium and the well developed theories of portfolio management and of asset pricing with a view to establishing whether there may be the possibility for a community of understanding between these theories and empirical method and the fundamental nature of property with all its peculiarities.

Finally the book concludes by summarising the nature of the issues and the views of a discriminating group of interested observers on the conclusions which can be drawn from the analysis carried out within each chapter.

The purpose of the volume will have been met if its contents and conclusions prove to be sufficiently robust to establish an enduring platform from which any future discussion can move forward.

The interests of property investment, and of those involved with it, will then have been well served.

REFERENCES

Bowie M:1982, 'Depreciation: Who hoodwinked whom?' EG 262 405

Darlow C and M Brett:1983, Introduction in Valuation and Investment Appraisal, Darlow C (ed), London, E.G. p.1-18

Debenham Tewson and Chinnocks:1985, 'Obsolescence: Its Effect on the Valuation of Property Investment', London, DTC

Fraser W D:1984, Principles of Property Investment and Pricing, London, Macmillan

Greenwell Montagu:1977, The Land Securities Investment Trust and the Implications for Property Valuation, London, GM

Hager P P and D J Lord:1985, 'The property market, property valuations and property performance measurement', Institute of Actuaries

Healey and Baker:1988, 'Quarterly Investment Report', March, London, H & B

Investment Property Databank:1986, The Property Index, London, IPD

Jones Lang Wootton:1987a, 'Obsolescence: The Financial Impact on Property Performance', London, JLW

Jones Lang Wootton:1987b, 'Property Index', Winter 1987, London, JLW

Management Anaysis Center (UK) Ltd:1985, 'Competition and the Chartered Surveyor: Changing Client Demand for the Services of the Chartered Surveyor, London, MAC

McIntosh A P J and S G Sykes:1985, A Guide to Institutional Property Investment, London, Macmillan

Rowe and Pitman:1982, 'Property Indices - Do They Make Sense? Property Research Issues No.10, October, London R & P

Salway P:1986, 'Depreciation of Commercial Property', CALUS, College of Estate Management, Reading

Sykes S G:1983, 'The uncertainties in property valuation and performance measurement', The Investment Analyst, No.5, 25-35.

Warburg Securities:1988, 'Property: Review of 1987 and Prospects for 1988', London, Warburg

CHAPTER 2

VALUATION TECHNIQUES: A MATTER OF EVIDENCE AND MOTIVE

1   Matching Method with Motive

At the outset, it is essential to distinguish between two valuation
exercises:

(a)   assessing the price at which a property would be likely to
      sell if offered for sale on the open market; that is market
      valuation, and;

(b)   assesing the inherent value of a property for the purpose of
      making investment decisions; investment appraisal

     The lack of a clear distinction by various commentators, e.g.
Sykes:1984, Trott:1980, has helped to prolong the debate over
valuation methods in the UK property profession.   Indeed Trott:1986
seems to compound the problem by confusing valuation practice with a
third aspect of the study of property valuation:

(c)   explaining how property values are determined by market
      forces; market price determination

     As the purposes of (a), (b) and (c) are distinct, it should not
be surprising if the analytical techniques are different.
     It is the objective of this paper to argue that the appropriate
valuation technique to adopt depends on the purpose of the valuation,
and furthermore in the case of market valuation, that the appropriate
technique can depend on the quality of sales evidence available for
comparable properties.
     A useful comparison can be made with the functions of a
stockbroker. Before making a buy/sell/hold recommendation, a
stockbroker's analyst will undertake a fundamental analysis. He will
investigate the likely future earnings (net profits) of the company
over the next two or three years and, after taking account of the
quality of the earnings and the risk of the company, he will make a
judgement about the inherent worth of the shares. If this is
significantly above the market value he will recommend a purchase, if
below he will tend to recommend a sale.
     On the other hand, if a client wishes to know the market value of
his shares, the broker will not undertake any analysis, but will call
up the appropriate page on his computer screen showing the current
trading price in the market.
     Perhaps this is a rather unfair analogy because, unlike the
homogeneous nature of shares in a single company, each property is
unique. A better comparison is with the stockbroker's function when
advising on the price at which to offer shares in a company going
public. Essentially he will analyse the price/earnings (P/E) ratios
of closely comparable companies and estimate the likely P/E of the
new shares when trading commences. He will then multiply this P/E
with the company's earnings to arrive at the predicted market price.

In assessing the market value of a share, the exercise is essentially one of objective comparison with market evidence of closely comparable investments. But when valuing for investment appraisal, the exercise is one of detailed analysis of risk and expected return, culminating in a subjective judgement as to the inherent value of the investment.

The same principle should hold for property valuation. Just as stockbrokers adjust their techniques according to the purpose of the valuation, the job of the valuer is to assess the value that the market considers the property to be worth. In the case of investment appraisal, the purpose is to assess what the valuer considers the property to be worth.

2 Valuation for Investment Appraisal

In property investment appraisal the valuer should make explicit allowance for risk and the principal variables affecting future return, particularly the prospects for rental growth and obsolescence. This implies a discounted cash flow (DCF) type of analysis.

Discounted cash flow is essentially a simple model enabling the present value of an investment to be calculated by discounting expected future income at a rate which reflects the 'quality' (in terms of risk, liquidity, expenses, etc.) of the investment.

The essential DCF concept can be expressed as follows:

$$V = \frac{R_1}{(1+r)} + \frac{R_2}{(1+r)^2} + \frac{R_3}{(1+r)^3} + \ldots \tag{1}$$

where V   = present value;
$R_{1,2,3}$ = expected rental income in periods 1,2,3, etc.
r   = investors' target (or required) return per period

Alternatively, the model can be expressed in more complex form to allow for rent review periods, to reflect explicitly the valuer's expectations for such variables as rental growth and depreciation, and to allow for the likelihood that rates of growth, obsolescence, and the target return will vary over time. In the simplifying case of a property recently let at full rental value, subject to annual rent review and with rent paid annually in arrears, the model could be stated as follows;

$$V = \frac{R}{(1+r_1)} + \frac{R(1+g_1)(1-d_1)}{(1+r_1)(1+r_2)} + \frac{R[(1+g_1)(1-d_1)][(1+g_2)(1-d_2)]}{(1+r_1)(1+r_2)(1+r_3)} + \ldots \tag{2}$$

Where R   = current rental value
$g_{1,2}$ = expected market rental growth rate in years 1 and 2
$d_{1,2}$ = expected rate of depreciation in years 1 and 2;
$r_{1,2,3}$ = investors' target return in years 1, 2 and 3

14

Another refinement which could be made to the above model would be to split r, the target return, into its two principal components, namely a risk free rate and a risk premium which reflected the qualities of the individual investment.

Valuation for investment appraisal therefore involves the valuer in making explicit judgements (based on historic evidence) about the property's future rental growth, rate of depreciation through obsolescence and risk. That will enable him to judge whether the property is appropriately valued in the market and, hence, enable him to give sound advice to his client on purchase or sale.

3 Models for Market Valuation

In investment appraisal the valuation is based on the valuer's opinion of risk, future growth, etc., whereas in market valuation it is the market's view about these variables that is relevant. Market values are based on the market's perception of, e.g. risk, future growth and obsolescence. It is irrelevant to market valuation whether the market is ignorant or irrational. In a relatively inefficient market like the property market, a valuer's opinion of a property's value may well be more sensible and rational than the market's but he can't arrive at a better market value than does the market itself.

Equation (2) above would be a valid model for market valuation as well as for investment appraisal if it was possible to quantify the market's perception of a property's risk, growth and obsolescence. But that cannot be done with reliability, and there are grave risks in ascribing to the market the valuer's own views on these variables.

As previously explained by the stockmarket analogy, the process of market valuation is essentially a process of comparison. In the majority of cases it is neither necessary nor desirable to undertake an analysis of the variables which determine value. The best evidence of the value of a subject property is the value of closely comparable properties as provided by sales in the open market.

As market valuation is a process of comparison, the most reliable valuation model will tend to be the one with the most reliable unit of comparison. Ideally such a unit of comparison should be readily identifiable from market evidence and reflect the source of value to the investor.

Setting aside 'positive' (see Wood:1985 and 1986) and 'real value' approaches (Crosby:1983) to market valuation, on the grounds that they seem to this author to add complexity to the valuation process without having compensating advantages, there are essentially two valuation models, years purchase (YP – equivalent yield) and discounted cash flow (DCF 'equated yield').

According to the YP concept, value is a multiple of the current rent (and rental value), the multiple being determined by investors' required yield. In the simple case of a rack rented freehold, the model is stated as follows:

$$V = \frac{R}{y} \tag{3}$$

Where V = market value of property
R = current net rent
y = investors' required income yield p.a. (all risks yield)

According to the DCF concept, a property's value is the present value of expected future rental income, the discount rate being investors' target rate of return (IRR) p.a. (see equation (1) above.) Alternatively, the DCF model may be restated in terms of the property's current rental value and its expected rental growth rate. The following equation represents a freehold property recently let at its rental value on a long lease subject to triennial rent reviews.

$$V = \frac{R}{(1+r)} + \frac{R}{(1+r)^2} + \frac{R}{(1+r)^3} + \frac{R(1+g)^3}{(1+r)^4} + \frac{R(1+g)^3}{(1+r)^5} + \frac{R(1+g)^3}{(1+r)^6} \tag{4}$$

$$+ \frac{R(1+g)^6}{(1+r)^7} + \cdots\cdots\cdots$$

Where R = current rental value
g = the property's expected rental growth rate p.a.

Equation (4) simplifies (see Fraser, 1986) to;

$$V = \frac{R}{r - r\left[\dfrac{(1+g)^3 - 1}{(1+r)^3 - 1}\right]} \tag{5}$$

and as R/V is the property's income yield (y), then the link between the YP and DCF models is shown by;

$$y = r - r \cdot \frac{(1+g)^n - 1}{(1+r)^n - 1} \tag{6}$$

where n = period between rent reviews (years)

Note. Whereas in equation (2) rental growth and depreciation are stated separately, g in equations (4), (5), (6) and (7) represents a rental growth rate which incorporates and allows for the market's expectation for future depreciation of the property through obsolescence.

Thus a property's yield (YP model) is determined by, and implicitly reflects, the property's rent review period, investors' target return and their expectations for the property's rental growth (and obsolescence). The essential difference between the two models

16

is that in the DCF model the variables are explicit and in the YP
model they are implicit (in the yield).

Both the concepts and models are flawless, and provided the
growth rate, target return and review period in the DCF model are
compatible with the yield adopted in the YP model, the solution will
be identical in each case.

4   The Practical Flaw of the DCF Method

On the assumption that the property to be valued is rack rented and
the rent is known, then in applying the YP model the valuer has only
one unknown to solve, the appropriate capitalisation rate, y. If
good comparable sales evidence is available, this can be identified
with sufficient accuracy to produce a reliable valuation. But in the
case of the DCF model there are essentially two unknowns, the
market's target return and expected rental growth rate, and neither
can be identified with absolute confidence.

The accuracy of market valuation depends crucially upon the
analysis of market evidence. Accordingly, when using the DCF method
for market valuation both the target return and rental growth rate
must be derived from the analysis of sales evidence. But in order to
identify the target return implied by a sale price, the valuer must
first quantify the market's rental growth expectation, but to
calculate the growth expectation the valuer must first estimate a
target return.

This requires clarification. The growth rate applied in market
valuation is the rental growth rate expected by the market. In order
to quantify this in an objective and practical way, it is necessary
to compare yields on rack rented freehold property (comparable to the
property being vaued) with an estimate of the market's target rate of
return for that property. In the case of good quality investment
property, this target rate of return is normally deemed to be 2%
above current yields on long dated 'gilt edged' stock. The growth
expectation can be calculted by using equation (7) below, which is
merely equation (6) restated in terms of g.

$$g = \left[\frac{(r-y)(1+r)^n + y}{r} \right]^{1/n} - 1 \qquad (7)$$

where g = the market's implied rental growth expectation (MIRGE)
            p.a.
      y = yield available on comparable rack rented freeholds
      r = estimated target return, rack rented freeholds

The one practical flaw in the above process is the inability to
identify precisely the target rate of return to be adopted in
equation (7). This problem arises because of the relative volatility
of gilt yields compared with property yields, and because the
validity of a 2% yield premium, or any yield premium, cannot be
proven. The appropriate yield premium must tend to vary between
different properties (e.g. due to differences in risk) and over time.
Indeed the author's research suggests that in the 1980s so far, the
market's target rate of return for good quality investment property

has varied over a range of approximately 4% relative to gilts, i.e. from a yield discount to gilts of about 2% in 1981, to a yield premium which might currently exceed 2% (Fraser:1986).

Although significant, this practical flaw does not invalidate the DCF process for market valuation. Errors will tend to be partly self-correcting in the sense that an excessively high target return applied in equation (7) will result in the application of an excessively high growth rate and target return in the valuation calculation. The one error will tend to cancel out the other and, in certain cases, the residual error could be negligible compared with an erroneous choice of capitalisation rate using the YP method.

5 The Valuation of Standard Freeholds

The practical flaw inherent in the DCF method of market valuation suggests that the YP method is to be preferred in conditions in which sales evidence from comparable property is of sufficient quality to enable the valuer to select reliably the appropriate capitalisation rate. The fact that no explicit allowance has been made for growth, obsolescence or risk is irrelevant, because with these variables being fully reflected in the market prices of the comparables, they must be allowed for implicitly in the selected capitalisation yield.

It is for valuers to judge in any particular case what constitutes suffiently good sales evidence, but its availability is unlikely to be a problem in the case of freehold investments with a standard pattern of income flow, due to the relative abundance of such property, i.e. let on standard leases with provision for regular rent reviews. However, as the risk and growth characteristics of a property (and the liability to tax) change marginally over the rent review cycle, one would expect market yields to vary accordingly. Just before a rent review the annual growth potential of an investment is higher than in the period just after a review. In the latter case, the investor has to wait several years before rental growth is converted into higher income, whereas in the former case the benefit is close at hand. One would therefore expect yields on reversions to be somewhat below those on rack rented property, but there is some evidence that the opposite can be the case (Sykes:1983). In either case, the valuer should be aware that yields may vary over the review period and, therefore, derive the capitalisation yield from comparables at a similar phase in the review cycle.

In the valuation of standard reversionary investments, it is important that the 'equivalent yield' variation of the YP method is adopted. This uses a single capitalisation rate to value both term and reversionary elements, in contrast to the traditional procedure in which the term income is normally valued at a rate somewhat below that of the reversion. The following is a simple example of a valuation of a freehold investment.

Example A freehold office investment has recently sold on the open market for £751,000. It is let on a long lease with a prime full repairing and insuring (FRI) covenant and rent reviews at five-yearly intervals. The current rent is £40,000, the rental value £50,000, and the next rent review is in two years time.

Current rent	40,000	
YP for 2 years @ 6.5%	1.821	72,840
Current rental value	50,000	
YP of rev. to perp. in		
2 years @ 6.5%	13.564	678,200
		£ 751,040

If the investment has sold for £751,000 then it reflects an 'equivalent' yield of 6.5%. But although that yield correctly values the investment as a whole, it does not correctly value either the term income or the reversion, considered in isolation. No rational investor would pay as much as £72,840 for an annual income of £40,000 for two years, when higher returns could be gained from buying a short dated gilt. The equivalent yield of 6.5% reflects a growth potential which the term income does not possess. That part of the investment is overvalued in the above calculation but, as the value of the whole investment is deemed to be correct, it follows that the reversion is similarly undervalued.

It is important to appreciate, therefore, that in times of rental growth the YP method does not purport to value the two individual parts of an investment accurately. Term and reversion are treated separately merely because their rents are different. The capitalisation yield adopted by the valuer must reflect the qualities of the investment as a whole. Therefore it seems anomalous to adopt the traditional practice of using a marginally lower yield to value the secure term income. It is incongruous to make minor adjustments of this type when no allowance is made for the lack of growth potential in the term income. It is also invalid to use a high cost of capital rate to value the fixed term income while valuing the reversion traditionally. That may produce a rational value for the term income but the reversion is now liable to be inappropriately valued. The only practical alternative is a full DCF valuation.

So, the term and reversion should be valued at the same yield which reflects the qualities of the investment as a whole, and which has been derived from the analysis of closely comparable sales evidence. If a single capitalisation yield is applied, it is worth noting that it is irrelevant whether the method illustrated above is ued or the alternative 'layer' or 'hardcore' method is adopted. The result must be identical.

Current rent	40,000	
YP in perp. @ 6.5%	15.385	615,400
Rental increase at reversion	10,000	
YP of rev. to perp. in		
2 years at 6.5%	13,564	
		135,640
		£ 751,040

6 The Valuation of Non-Standard Investments

The multiple by which capital value exceeds rent (and rental value)
is determined by the capitalisation yield. There must always be a
yield which provides the 'correct' multiplier, and if market evidence
is sufficient to identify this yield accurately then the YP method is
valid and sufficient for market valuation purposes. However, in
cases such as high-geared or short leaseholds, complex equity-sharing
interests, certain reversionary freeholds and other investments
having a non-standard pattern of income flow, good sales evidence is
unlikely to be available. Furthermore, in these cases the growth and
risk characteristics of the investment can be complex and obscure,
rendering the choice of capitalisation rate particularly prone to
error.

The principal determinants of a property's yield are risk and
growth. Indeed in the case of property of a quality acceptable to
institutional investors, growth expectation is probably the dominant
consideration. But whereas the capital growth of a standard freehold
will relate closely to the underlying rate of rental value growth, a
common feature of non-standard property investments is that their
capital growth can vary substantially from the underlying rental
value growth.

Certain reversionary freeholds can possess individualistic growth
characteristics, particularly in cases where long periods elapse
between rent reviews. The capital growth of such investments will
depend not only upon the underlying rental growth of the property,
but also upon the proximity of the reversion and the ratio of the
current rent to the rental value. In cases where the next review is
beyond, say seven years, or where the current rent is a small
proportion of rental value, growth characteristics will be
particularly difficult to gauge and good comparable evidence is
likely to be scarce. Due to the difficulty of identifying the
appropriate capitalisation rate in these cases, a DCF calculation is
recommended, at least as a check.

The problem of obtaining sales evidence of good comparable
property is likely to be even greater in the case of leasehold
investments. Inevitably, leaseholds are more disparate than
freeholds, partly because of their finite duration but also because
their investment characteristics are determined by the conditions of
the head-lease as well as the sub-lease.

The valuation problem is exacerbated, particularly in the case of
short leaseholds, by the complexity of their growth. Whereas in the
case of freeholds, rental growth will ultimately be converted into

20

capital growth, the short leasehold investor may be benefiting from income growth (through rent reviews on the sub-lease) while at the same time suffering capital loss as the head-lease nears its end. The growth and risk characteristics may be further complicated by a variable head-rent or by an element of income gearing (where a fixed head rent is a significant proportion of the sub-rental income).

The combined effect of differing lease duration, review period, gearing, etc., not to mention the covenant and physical attributes of a property, can make the selection of capitalisation rate, by reference to an apparently comparable investment, peculiarly prone to error. On the other hand, the ability of the DCF approach to strip out and quantify the market's implied future rental income expectation, taking account of lease duration, review period, gearing, etc., leaves the valuer with the relatively simple task of selecting an appropriate discount rate. The DCF method is more suited to the valuer's function of 'fine tuning' and, of course, it avoids the various anomalies and weaknesses of the traditional dual rate YP system (see Fraser:1977). It seems that this method has now been largely discredited by practitioners.

7   Conclusion

This paper has been concerned with the methods and techniques which the author feels should be adopted for the valuation and appraisal of property investments. The essential moral is that the optimal method in any particular case depends on the purpose of the valuation and the market evidence available. If the objective is to carry out an appraisal for investment decision-making, the valuation will be the valuer's own view of the property's worth, and should reflect his views on such variables as rental growth, obsolescence, and risk. But if the objective is market valuation, then the personal views of the valuer on such matters are irrelevant. The valuation exercise is essentially one of objective comparison and only those variables which have been quantified from comparable sales evidence should be incorporated. If sufficient evidence of close comparables is available to enable the capitalisation rate to be reliably identified, the YP method is considered optimal, but in the cases of leaseholds and other investments with a non-standard pattern of income flow, the DCF method will probably tend to be the more reliable.

However, the choice of method is not mutually exclusive. In the case of the market valuation of non-standard investments, a DCF should normally be backed up by a conventional valuation. The reason for any disparity between the two results should be investigated and the valuer's judgement used to reconcile the difference.

Finally, it is important to avoid an irrational concern about the theoretical accuracy of valuing an income flow, while ignoring the variability that must inevitably result from the many imponderable variations in lease covenants and the locational and physical qualities of property. A valuation is an expert judgement and the calculation is merely a means to that end.

21

References

Crosby N:1983, The Investment Method of Valuation: A Real Value
    Approach, Journal of Valuation 1: 341–350, and 2:48–59.
Fraser W D:1977, The Valuation and Analysis of Leasehold Investments
    in Times of Inflation, Estates Gazette 244: 197–203.
Fraser W D:1986, Property Yield Trends in a Fluctuating Economy,
    Journal of Valuation 4: 239–260.
Sykes S:1983, Valuation Models: Action or Reaction?, Estates
    Gazette, 267: 1108–1112.
Sykes S:1984, Property Valuation: A Rational Model, Journal of
    Valuation 2: 258–270.
Trott A J:1980, Property Valuation Methods: Interim Report,
    Polytechnic of the South Bank/RICS.
Trott A J:1986, The Use of Conventional Valuation Methods in Times of
    Inflation, in Trott A J (ed), Property Valuation Methods:
    Research Report, Polytechnic of the South Bank/RICS.
Wood E:1985 and 1986, Positive Valuation Methods 1 and 2, Journal of
    Valuation 4: 7–15 and 170–184.

CHAPTER 3

THE CURRENT DISCOUNTED CASH FLOW MODELS FOR THE VALUATION AND
ANALYSIS OF PROPERTY INVESTMENTS: AN EXAMINATION OF SOME OF THE
PROBLEMS

1. Introduction

Methods of property investment valuation and analysis have been
subject to much criticism over the past decade. Indeed, the scrutiny
of conventional methods has its origin in the research carried out by
Greaves:1972 and Wood:1972. Both Greaves and Wood were very critical
of the conventional methods and called for the use of discounted cash
flow models in property valuation and analysis, but the response from
the property profession has been very disappointing.

Recently two DCF models have been formulated for the valuation of
property investments. They are the Rational Model (McIntosh and
Sykes:1982) and the Real Value Approach (Crosby:1983). These models
are based on similar assumptions but in their applications they take
slightly different forms. The applications of these models for the
valuation of freehold reversions and leaseholds have been reviewed by
Baum:1984, and Baum and Yu:1985.

The purpose of this chapter is to examine some of the problems
that are involved in the use of contemporary discounted cash flow
models for property investment valuation and analysis. This chapter
examines the application of these models, the assumptions on which
they are based and their limitations in certain situations. The
equated yield version of the DCF model is considered in this chapter,
but the comments should be equally valid for the Rational Model and
the Real Value Approach.

2 Discounted cash flow models

The discounted cash flow models can be used for both valuations and
analysis (or appraisal). The purpose of a valuation is to estimate
the market price of an investment in property.   An appraisal (or
analysis) is usually undertaken to estimate the investment worth of a
property investment to a particular individual, based on his
estimates of rental growth, yields and risk attached to the
investment.

It is fair to say that the majority of valuations are based on
conventional methods and the use of explicit DCF models appears to be
confined to investment appraisals. However, there is some evidence
to suggest that DCF models are used for the valuation of short
leasehold investments (Baum:1986).

The application of DCF models for the valuation of rack rented
(fully let) freehold investments, reversionary freehold investments
and leasehold investments are discussed below.

Rack rented freehold investments

The capital value of a rack rented freehold investment may be
expressed by the following equation:

$$V = R.YP(n) + R(1+g)^n.YP(n).(1+r)^{-n} + \ldots$$

$$+ R(1+g)^{mn}.YP(n).(1+r)^{-mn} + \ldots \tag{1}$$

Where

R = current full rental value
YP(n) = present value of an annuity or year purchase for n years
    at discount rate r
n = rent review period
g = market's expectation of annual growth rate in rental
    values (assumed constant)
r = the risk adjusted discount rate (or equated yield)

Equation (1) represents the discounted present values of the income flows from an investment in perpetuity.

The value can also be expressed in the following form

$$V = \frac{R. \quad YP\ (n)}{1 - \left[\dfrac{1+g}{1+r}\right]^n} \tag{2}$$

In deriving the above equation it is assumed that the income is received at the end of each year.

It will be evident from (2), that the value V can be determined only if R, n, g and r are known. The values of g and r are not directly observable for property investments and have to be proxied. It is for this reason that no serious attempt is made to value rack rented freeholds let on standard rent review patterns using a DCF approach. Instead the sale price of such an interest is used as a point of reference for the valuation of freehold reversions and leaseholds.

Referring to equation (2) the value of the rack rented freehold assuming no growth in rental values can be expressed as

$$V = R.\frac{1}{y} \text{ , where } y = \text{the discount rate, when } g = 0.,$$
     known as the aLl risks yield

$$\frac{1}{y} = \frac{YP\ (n)}{1 - \left[\dfrac{1+g}{1+r}\right]^n} \tag{3}$$

Equation (3) depicts the relationship between the variables k, r and g.

If the market's target rate (equated yield) can be proxied by reference to redemption yields on long dated gilts plus a premium for risk, liquidity, (usually 2%), then the value of g, the market implied growth rate expectations, can be determined using equation (3), given that the all risks yield y can be derived directly from market evidence. Thus if the all risks yield y = 5.00%; the review period n = 5; and the discount rate r = 11%, then the implied growth rate g = 6.5556% (See Appendix 1).

It should be noted that the derivation of the implied growth rate is the starting point for the valuation of reversions and leases and the value of g thus derived is used in both valuations.

Reversionary freehold investments

The value of a freehold reversion can be expressed by the following equation.

$$V = R(t).YP(T) + R\left[\frac{1+g}{1+r}\right]^t . \frac{YP(n)}{1 - \left[\frac{1+g}{1+r}\right]^n} \qquad (4)$$

or

$$V = R(t).YP(T) + R\left[\frac{1+g}{1+r}\right]^t . \frac{1}{y} \qquad (5)$$

Where
R(T) = rent during term
R = current full rental value
g = implied growth rate assumed constant
r = the discount rate
t = length of the term
y = the all risks yield for similar rack rented freehold
YP(T) = PV of an annuity for T years (years purchase)
YP(n) = PV of an annuity for n years (years purchase)

The above DCF model assume that the yield (the all risk yield) of rack rented freehold remains unchanged throughout the life of the investment.

The DCF model represented by equation (5) can be illustrated by the following example.

Value the freehold reversion in a small shop let at a rent of £10,000 pa. There will be a rent review in 3 years time. The current full rental value is £15,000 pa.

Assume the all risks yield = 5%; (from analysis of comparable sales)
discount (equated yield) rate = 11%; implied growth rate = 6.5556% from (3)

Valuation 1

Term rent		£10,000 pa	
YP 3 years @ 11%		2.4437	
			£24,437
Reversion rent	15,000(1.06556)^3	£18,148pa	
YP in perp. @ 5%	20.00		
PV £1 in 3 years @ 11%	0.7312	14.6238	
			£265,393
			£289,830

25

(i) The same discount rate of 11% is used to value the reversion, implying that the market views the rack rented freehold and the reversion as equally risky.

(ii) If $g = 0$, then equation (4) represents a conventional equivalent yield valuation where r is the equivalent yield. Thus using equation (4) it can be shown that an equivalent yield of 4.943% will lead to the same value.

Leasehold investments

The value of a leasehold investment with a fixed head rent for the term can be expressed as follows.

$$V = (R-R(H)).YP(n)+(R(1+g)^n-R(H)).YP(n)(1+r)^{-n}+...$$

$$+(R(1+g)^{mn}-R(H)).YP(n).(1+r)^{-mn} \quad (6)$$

$$= R.YP(n)\left\{\frac{1 - \left[\frac{1+g}{1+r}\right]^{(m+1)n}}{1 - \left[\frac{1+g}{1+r}\right]^n}\right\} - R(H).YP[(m+1)n] \quad (7)$$

Where

R(H)	=	rent payable (fixed) for the term
R	=	rent paid by subtenant subject to review every n years
n	=	rent review period
(m+1)n	=	unexpired term of the lease
YP(n)	=	PV of an annuity for n years (years purchase)
YP(m+1)n=		PV of an annuity for (m+1)n years or length of the lease
r	=	discount rate (leasehold equated yield)

The use of equation (6) is demonstrated by the following example.

Example

A leasehold interest has 15 years to run at a fixed rent of £10,000 pa on FRI terms. The current full rental value is £15,000 pa on FRI terms. The property is sublet to the occupying tenant on 5 yearly reviews.

It is necessary to reflect the market's expectation of future rental growth. In the current DCF models, the market's implied growth rate in rentals is derived from an analysis of a similar rack rented freehold and the resulting figure is used for the valuation of the lease.

Thus if y = A.R.Y = 5%
         r = 11%
   then g = 6.5556%   (from 3)

The valuation can be set out as follows:

Valuation 2

i	Rent received	£15,000	
	<u>less</u> rent paid	£10,000	
	Profit rent	£ 5,000	
	YP 5 years @ 16%	3.2743	
			£16,371

ii	Estimated full rental value $15,000(1.065556)^5$	£20,605	
	<u>less</u> rent paid	£10,000	
	Profit rent	£10,605	
	YP 5 years @ 16% 3.2743		
	PV 5 years @ 16% <u>0.4761</u>	1.5589	
			£16,532

iii	Estimated full rental value $15,000(1.065556)^{10}$	£28,304	
	<u>Less</u> rent paid	£10,000	
	Profit rent	£18,304	
	YP 5 years @ 16% 3.2743		
	PV 10 years @ 16% <u>0.2267</u>	0.7423	
			£13,587
			£46,490

Notes

(i)   A discount rate of 16% (5% higher than the freehold rate of
11%) was used to reflect the higher risks associated with a
geared lease with 15 years to run.   The choice of this rate is
purely subjective.

(ii) The DCF model can also be adapted for leases where the head
rent is variable and the lease term is not an exact multiple of
the review period.

So far, the two examples illustrate the application of contemporary DCF models for the valuation of freehold reversions and
leaseholds.   The assumptions behind these models are important as
they will determine the usefulness and validity of the models in a
particular situation. The main assumptions are considered in the
following sections.

3   The Assumptions

The discount rate is an important variable in the valuation of
freehold reversions and leaseholds and this is considered first
followed by a consideration of the assumption relating to the growth
rate.

The discount rate

The current DCF models assume a single rate of discount which discounts the income flow for each year. As noted earlier, the estimation of the discount rate is a prerequisite for the derivation of the implied growth rate. The discount rate is usually derived by reference to the redemption yields from government bonds of similar maturity. The models assume annual receipt of rent in arrears. However, rents are paid quarterly in advance, whereas interest payments from bonds are paid half yearly in arrears. The redemption yield is dependent on the date of purchase, usually lying between interest dates, the magnitude of the coupon payments and the term to maturity. Thus there is no one-to-one correspondence between the redemption yield and maturity (Abayagunawardana:1980, Morely:1983). In the use of the DCF models the discount rate is derived by adding a subjective risk premium (usually 2%) to the redemption yield on long dated government bonds. In view of the subjective nature of the risk premium, detailed attention to the coupon, tax and maturity effects are not normally paid. The rate of return that investors do require and hence the implied growth rate can not be determined precisely.

Clearly, the implied growth is an important variable for investment analysis and valuations. However, it can be shown that variations in the value of discount rate will not significantly affect the capital values of freehold reversions derived from the DCF models in most cases (Crosby:1986). In the analysis of rack rented freeholds, high (low) discount rates lead to high (low) implied growth rates and when these values are used for the valuation of reversions, they tend to cancel each other leaving the capital values virtually unaffected. Capital values of leasehold investments are, however, more sensitive to growth rates due to the effects of gearing and shorter lives of such investments and in these cases the selection of the discount rate will have a significant impact on values.

Capital market theory and the discount rate

The rate of return that the market participants do require cannot be known precisely, but in the use of DCF models for valuations it is deemed to be two per cent above the redemption yields on gilts. However, for investment appraisals, the universal use of a two per cent risk premium is inappropriate as this would imply that all investments belong to the same risk class.

Attempts have been made to derive the expected rate of return for property using the Capital Asset Pricing Model (Ward:1981, Brown:1984, Fraser:1986). Brown:1984 has estimated the market risk for different sectors of the property market as follows:

Retail 0.23
Offices 0.17
Industrials 0.14

The above figures imply a higher expected rate for retail followed by offices and industrials. These are very useful develcpments but more research needs to be done before reliable estimates of market risk for individual properties can be made.

Keane:1977 describes the difficulties that are present in the derivation of a risk adjusted discount rate as follows:

'The discount rate for a specific project cannot be derived from the "firm's costs of capital" or from the 'market rate of return' both of which are average rates for investments of varying maturities and risks. It is necessary to refer to the rate on a publicly traded asset of like maturity and risk, and this effectively implies reference to the corporate bond market.'

It is thus evident that even from the point of view of appraisal, proper risk adjusted discount rates are not easily obtainable.

The foregoing discussion outlined, some of the problems involved in the selection of an appropriate discount rate. However, the discount rate here is taken as the redemption yield on long dated gilts plus a premium.

Valuation of freehold reversions

It is appropriate to mention that in the valuation of freehold reversions, the contemporary DCF models use the same discount rate that was used for the analysis of rack rented freeholds. This assumption implies that the discount rate is independent of the lease structure of the reversion (Baum:1984; Fraser:1985). The following tables show the equivalent yields required to produce the same capital values derived from DCF valuations of varying lease structures.

In Table 1, it is assumed that the all risks yield of rack rented freehold in identical property (shop) is 5% and the discount rate (equated yield) is 11%, giving an implied growth rate of 6.5556%.

In Table 2, it is assumed that the all risks yield of rack rented freehold in identical property (office) is 9% and the discount rate is 14%, giving an implied growth rate of 5.8774%.

Table 1   Conventional equivalent valuation to equate with DCF approach assuming all risks yield = 5%, discount rate = 11%

Equivalent Yield

Size of Reversion			Length of Term	
Term Rent:FRV	5	10	15	20
3:4	4.96	5.09	5.33	5.63
2:3	4.95	5.05	5.25	5.52
1:2	4.92	4.97	5.11	5.29
1:4	4.88	4.86	4.88	4.95

Table 2   Conventional equivalent valuation to equate with DCF
approach assuming all risks yield = 9%, discount rate = 14%

Equivalent Yield

Size of Reversion		Length of Term		
Term Rent:FRV	5	10	15	20
3:4	8.91	9.25	9.81	10.99
2:3	8.88	9.17	9.67	10.27
1:2	8.81	8.99	9.36	9.86
1:4	8.71	8.71	8.85	9.10

The above figures show that the equivalent yield varies directly
with the length of time to reversion but inversely with the size of
reversion. It will be extremely difficult to test empirically if the
DCF approach gives a better guide to market values as it will not be
possible to find sales of reversions of varying lease structures at a
given point in time. However, market sentiment might dictate the use
of higher equivalent yields than those shown in these tables.

If we ignore the complications caused by taxation and assume that
the market is likely to use higher equivalent yields, then the values
derived from DCF methods may not lead to an accurate assessment of
market values. Table 3 gives the discount rates that should be used
to equate with values obtained on the assumption of the following
market based equivalent yields.

Table 3   Conventional equivalent valuation to equate with DCF
approach assuming all risks yield = 5%, discount rate = 11%
           growth rate = 6.5556%;

Term rent:FRV = 2:3

Term	Equivalent Yield	DCF Rate
5	5.00	11.05
	5.25	11.28
10	5.50	11.39
	5.75	11.61
15	6.00	11.61
	6.25	11.81
20	6.75	11.92
	7.00	12.11

If a market valuation is sought, then evidence of transactions
needs to be analysed and if they indicate higher equivalent yields as
shown above, the DCF approach has to be used accordingly with higher
values for discount rates.

From the view point of investment appraisal, it is therefore tempting to arrive at the conclusion that the market undervalues reversionary investments. However, we cannot overlook the argument that reversions with varying lease structures might warrant the use of different discount rates commensurate with their levels of risk. A simple but crude means of testing if this should be the case, will be to consider the variations in capital values of reversions with different lease structures for variations in discount rates and implied growth rates. The results of the sensitivity tests are presented in Appendix 2, at the end of this chapter.

Sensitivity of capital values of reversions

The following observations can be made from the figures in Appendix 2 at the end of this chapter.

(a)   Values of reversions vary inversely with discount rates.
(b)   For a given change in the discount rate, the percentage changes in the values are greater, the longer the period to reversion, over the periods considered. (However, beyond a certain period to reversion the percentage changes begin to decline.)
(c)   The higher the ratio of the rent during term to the current full rental value, the smaller will be the percentage value fluctuation for a given percentage change in the discount rate.
(d)   The percentage value changes increase at a diminishing rate as the length of the period to reversion increases.

The above observations are made on the basis of results obtained from a few examples. A general derivation of these properties will be required if they are to hold for all possible ranges of values. These variations are similar to bond price movements for changes in interest rates. The redemption yields of bonds vary with the coupon and the term to maturity. It is, therefore, intuitively appealing to suggest that freehold reversions should be valued at different discount rates depending on their lease structures and levels of risk. This point has not been seriously considered, but it is worthy of more attention.

Valuation of leasehold investments

The selection of an approximate discount rate for leasehold investments is even more problematic as leases vary largely in their lease structures and evidence of comparable transactions not easily available. Leasehold investments are considered as more risky than freeholds due to the top-slice nature of the income and due to their shorter terms.

In the application of contemporary DCF models, a higher discount rate is therefore used to reflect the higher level of risk associated with leasehold investments. However, there are no established criteria for the selection of this rate. In general, leases are valued on a dual rate basis and therefore it would appear that a very

31

high discount rate is used in DCF valuations to ensure that valuations produced by DCF techniques are not totally out of line with conventional dual rate valuations.

Analysis of leasehold investment transactions may be useful in the choice of a discount rate. However, there are difficulties with this approach as leases tend to be unique in terms of lease structures and the quality of tenant's covenants. Further, the market's expectations on rental growth (which cannot be determined precisely) is required to determine the rate of discount, and these limitations need to be recognised in the valuations.

The implied growth rate

A major problem with the use of explicit DCF models concerns the forecasting of future rental values. We know very little about the factors that determine future rental growth or about the process by which future growth expectations are formed by investors. Since growth rate expectations are not directly observable, they have to be proxied. The constant implied annual growth rate derived from the analysis of freeholds may be regarded as a discounted average of annual rental growth into perpetuity (Fraser:1986). It could be possible that the investors' expectations may be dependent on the time horizon over which expectations are held (Malkiel and Cragg:1970, Van Horne:1978). The implied growth rate is a long term average rate, but due to the cyclical nature of the property market short term growth rates can be significantly different. Under these conditions, even if the average growth rate is achieved over the long term, it might not be achieved at specific rent review dates (Morley:1983). The implications of possible divergence between the short term rate and the long term average rate for valuations may be examined by the use of a two growth rate model.

A two growth rate model

The two growth rate model assumes growth expectations until the first review at a short term growth rate (g_1) and at a constant growth rate (g_2) thereafter into perpetuity. In analysing the rack rented freehold, a short term growth rate (g_1) is imposed up to the first review and the constant implied growth rate thereafter (g_2) is calculated.

The relationship between the implied annual growth rate g, the short term growth rate g_1 and the constant implied growth rate thereafter g_2 is given by the following equation (see Appendix 3).

$$g_2 = \left[\left\{ 1 + \left[\frac{1+g_1}{1+r}\right]^n - \left[\frac{1+g_1}{1+g}\right]^n \right\} (1+r)^n \right]^{1/n} - 1$$

Values of g_2 for assumed values of the all risk yield y, discount rate r, and short term growth rate g_1, are given below.

Table 4

ARY(y)	r	n	g	g_1	g_2
5.00	11.00	5	6.5556	10.00	5.7094
5.00	11.00	5	6.5556	12.50	5.005
9.00	14.00	5	5.8774	10.00	3.8053
9.00	14.00	5	5.8774	12.50	2.2864
5.00	11.00	5	6.5556	3.00	7.3000

From the above table, we may say that an implied annual growth rate of 6.5556% is equivalent to an implied growth rate of 10% for the first five years, followed by a constant implied rate of 5.7094% into perpetuity.

The following example shows the valuation of a reversion (see Valuation (1)) using the two growth model.

Freehold reversion

Valuation (3)

Term	= 3 years	Rent paid	= £10,000 pa
Current FRV	= £15,000 pa	Discount rate	= 11.00%
Growth rate g	= 6.5556%	Short term growth	
Growth rate thereafter g_2	= 5.7094%	rate g_1	= 10.00%

Value of reversion at constant growth = 289,830 (see Valuation 1)

Valuation using two growth rates

Term rent		£10,000	
YP 3 years at 11%		2.4437	24,437

Reversion rent			
$15,000(1.10)^3$		£19,965	
YP 5 years @ 11%	3.6959		
PV 3 years @ 11%	0.7312	2.7024	53,953

Reversion rent			
$15,000(1.10)^5 (1.057094)^3$		£28,536	
YP in perp. @ 5.8620%*	17.0589		
PV 8 years @ 11%	0.4339	7.4019	211,221
			289,611

* From equation (3)

33

The difference in values in this case is negligible.

In Appendix 4, percentage variations in capital values derived from figures in Table 4, for reversions of different lease structures are given. It can be seen from Appendix 4, Table A, that there is no significant error (10% or more) except for reversions greater than 15 years. Even here the short term growth rate (12.5%) has to be much higher than the implied average rate of 6.5556%.

In Table B of Appendix 4, the variations are relatively higher, but they relate to an old property and such high growth rates in the short term may be unlikely. Thus, it may be concluded that the values of reversions derived by the use of a single implied annual growth rate are not significantly affected by high short term growth rates except in cases where the term to reversion is 15 or more years. It will be important to examine whether the same conclusion applies to leasehold interests. The following example is considered for purposes of illustration.

Leasehold interest
(Valuation 4)

Head rent (fixed)	= £10,000
Current full rental value	= £15,000
Comparable rack rented yield	= 5%
Freehold discount rate	= 11%
Implied growth rate g	= 6.5556%
Let the short term growth over the next 5 years g_1	= 3%
Implied growth rate thereafter g_2	= 7.3%
(from equation 8)	
Leasehold discount rate	= 16%

The cash flows from leases of various durations are as follows:

Period	ERV	Rent Paid	Profit Paid	YP at 16%	PV at 16%	Value
(a) Constant implied growth			g = 6.5556%			
1-5	15,000	10,000	5,000	3.2743	1.000	16,372
6-10	20,605	10,000	10,605	3.2743	0.4761	16,533
11-15	28,304	10,000	18,304	3.2743	0.2267	13,585
16-20	38,880	10,000	28,880	3.2743	0.1079	10,206
21-25	53,409	10,000	43,409	3.2743	0.0514	7,304
(b) Variable growth rates g_1 = 3% g_2 = 7.3%						
1-5	15,000	10,000	5,000	3.2743	1.000	16,372
6-10	17,389	10,000	7,389	3.2743	0.4761	11,519
11-15	24,733	10,000	14,733	3.2743	0.2267	10,936
16-20	35,178	10,000	25,178	3.2743	0.1079	8,895
21-25	50,034	10,000	40,034	3.2743	0.0514	6,736

From the above figures we can summarise the results as follows:

(1) Length of Lease	(2) Value at Constant Growth Rate	(3) Value at Variable Growth Rates	% Difference
5	16,372	16,372	0.00
10	32,905	27,891	17.98
15	46,490	38,827	19.74
20	56,696	47,722	18.80
25	64,000	54,458	17.52

The above figures demonstrate the inherent dangers in using a long term average implied growth rate to value leases of shorter duration, when short term growth expectations are likely to be significantly different from the implied growth rate.

Holding period of the investment

In valuing freehold investments it is assumed that the income from the property is receivable in perpetuity. The contemporary DCF models, in addition to this assumption, further assume constant growth in rental values. These assumptions imply that if the investment is held for a finite period the yield (all risks yield) remains the same throughout the period.

In reality the yields of property investments change through time due to economic factors and depreciation. It is therefore evident that any analysis of rack rented freeholds for implied growth rate expectations is of limited value if the investor's holding period is finite. In such cases the yield applicable to the investment at the end of the period is likely to be different. Depreciation affects rental growth and the value of the investment at the end of the holding period through changes in yields (Salway:1986). The current models will therefore underestimate the implied growth rate for investments which are prone to depreciation.

A more realistic value for the implied growth can be determined by the use of finite DCF models. The growth rate (g) in the DCF model may be regarded as an annual growth rate net of depreciation. The relationship between the growth rate for new property and the growth rate for an old building can be expressed by the following equation

$$1+g = \frac{1+g_m}{1+d} \qquad\qquad (9)$$

where g = growth rate for subject property
d = annual rate by which depreciation affects rental growth
g_m = growth rate for new building

35

The following example illustrates the limitations of an infinite holding period assumption, in determining the implied growth rate.

Rack rented freehold investment (new building)
Purchase price   = £1,000,000
Full rental value   = £50,000
Holding period   = 20 years
Discount rate   = 11% pa
Rate of depreciation   = 3% pa
in rental values
(assumed constant)
Value of investment   = 35% of value of new building
at end of 20 years
Rent review period   = 5 years

(a) DCF analysis assuming an infinite holding period

All risks yield y = $\dfrac{50,000}{1,000,000}$ x 100 = 5%

Discount rate r = 11%

Implied growth rate g = 6.5556% (see Appendix 1)

(b) DCF analysis assuming holding period of 20 years.   Let g be the implied growth rate (net of depreciation)

Let g_m be the growth rate for new buildings

    $1 + g_m = (1+g)(1+d)$

Value of building at the end of 20 years

$= 1,000,000(1+g)^{20}(1+d)^{20}$ x 0.35 $= 1,000,000(1+g)^{20}(1.03)^{20}$ x 0.35

$\therefore$ 1,000,000 = 50,000 x YP 5 years at 11% $\dfrac{1 - \left(\dfrac{1+g}{1.11}\right)^{20}}{1 - \left(\dfrac{1+g}{1.11}\right)^{5}}$

             $+ 1,000,000(1+g)^{20}(1.03)^{20}$ x 0.35 x $\dfrac{1}{(1.11)^{20}}$

$\therefore$ $g = 8.2392\%$

The value of the growth rate with a finite holding period is 8.2392%. (Compare the rate of 8.2392% with 6.5556% obtained from an infinite DCF model.) In the above example, the difference of 1.6836% is due to the fact that the subject property is assumed to have a value of only 35% of the value of a new building after 20 years. The difference between the two rates will be dependent on the ratio of the value of subject property at the end of the period to the value of a new property at that point in time. The lower the value of this ratio, the greater will be the difference between the two rates.

The above example illustrates the limitations of an infinite holding period assumption. (Or the assumption that yields remain the same) for purposes of investment analysis. In valuations, the two different holding periods are not likely to produce significantly different values for reversions. However, with leases, it is possible to have different values if valued on the basis of growth rates derived from freeholds under the assumption of finite and infinite holding periods. It should be of interest to note that the differences in values for reversions based on finite and infinite holding period assumptions will not be significantly different but the values are derived on different growth rate expectations.

In finite DCF models it is possible to allow for the full impact of depreciation. They also have advantages in risk explicit appraisals. However, they are more complex requiring the estimation of more variables and are potentially dangerous if no proper allowance is made for the uncertain nature of variables such as rates of depreciation, terminal value of the investment and rental growth rates

4 Conclusions

This chapter has been concerned with the application of DCF models for the valuation and analysis of property investments. While the current models are shown to be adequate for the valuation of standard freehold reversions, it is argued that the application in their present form, for the valuation of distant reversions and geared short leases, requires further examination and refinement. Firstly, their reliance on the use of the same discount rate for the rack rented freehold and the reversion is not totally acceptable in view of the differences in lease structures. Secondly, the use of a long term implied growth rate in valuing geared short leasehold investments can result in distortions in valuations whenever short term growth rate expectations and the implied growth rate are likely to be significantly different.

Finally, it is argued that the present models tend to under-estimate the implied growth rate expectations of investments which are highly prone to depreciation.

Clearly, more research is required in the areas of risk analysis and forecasting if explicit cash flow models are to be firmly established in property investment valuation and analysis.

References

Abayagunawardana D P P:1980, 'Individual Worth Approach', Estates
    Gazette, Vol.256. pp 51-53
Baum A:1984, 'The Valuation of Reversionary Freeholds: A Review',
    Journal of Valuation, Vol.3 pp.158-69
Baum A, and Yu Shi Ming:1985, 'The Valuation of Leaseholds; A
    Review', Journal of Valuation, Vol.3, pp.157-67 and pp.230-247
Baum A:1986, 'The Valuation of Short Leaseholds for Investment',
    Property Valuation Methods: Research Report, Polytechnic of the
    South Bank/RICS, pp.59-78
Brown G:1984, 'Assessing an all risks yield', Estates Gazette,
    Vol.269, pp.700-6
Crosby N:1983, 'The Investment Method of Valuation: A Real Value
    Approach', Journal of Valuation, Vol.1 pp.341-50 and Vol.2,
    pp.48-59
Crosby N:1986, 'The Application of Equated Yield and Real Value
    Approaches to Market Valuation', Journal of Valuation, Vol.4
    pp.158-69, and pp.261-74
Fraser W:1985a, 'Rational Models or Practical Methods', Journal of
    Valuation, Vol.3, 353-358
Fraser W:1985b, 'The Target Return on UK Property Investments',
    Journal of Valuation, Vol.4, pp.119-29
Fraser W:1986, 'Property Yield Trends in a Fluctuating Economy',
    Journal of Valuation, Vol.4 pp.239-60
Greaves M J:1972, 'The Investment Method of Property Valuation and
    Analysis - An Examination of Some of its Problems', Unpublished
    PhD thesis, University of Reading.
Keane S M:1977, 'The Irrelevance of the Firm's Cost of Capital as an
    Investment Decision Tool', Journal of Business Finance and
    Accounting, Vol.1:2 pp.201-16
McIntosh A and Sykes S:1982, 'Towards a Standard Property Income
    Valuation Model - Rationalisation of Stagnation?', Journal of
    Valuation, Vol.1 pp.117-25
Malkiel B and Cragg J:1970, 'Expectations and the Structure of Share
    Prices', American Economic Review, September pp.601-617
Morley S:1983, 'Investment Appraisal', in Darlow C (ed), Valuation
    and Investment Appraisal, Estates Gazette, pp.287-319
Salway F:1986, Depreciation of Commercial Property, Research Report,
    CALUS
Van Horne J C:1978, Financial Market Rates and Flows, Prentice-Hall
    Inc., Englewood Cliffs, New Jersey, pp.126-128
Ward C:1981, 'The Evaluation of Risk', Estates Gazette, Vol.260
    pp.253-56
Wood E:1972, 'Property Investment - A Real Value Approach'
    unpublished PhD Thesis, University of Reading

Appendix 1

From equation (3)

$$\frac{1}{y} = \frac{YP(n)}{1 - \left[\frac{1+g}{1+r}\right]^n}$$

$$1 - \left[\frac{1+g}{1+r}\right]^n = y.YP(n)$$

$$\left[\frac{1+g}{1+r}\right]^n = 1 - y.YP(n)$$

$$1+g^n = (1+r)^n - y.\left[\frac{1 - (1+r)^{-n}}{r}\right] . (1+r)^n$$

$$= \frac{r(1+r)^n - y(1+r)^n + y}{r}$$

$$g = \left[\frac{(r-y)(1+r)^n + y}{r}\right]^{1/n} - 1$$

$r = 11\%$, $y = 5\%$, $n = 5$

$$g = \left[\frac{(0.11 + 0.05)(1.11)^5 + 0.5}{0.11}\right]^{1/5} - 1$$

$$= 0.65556$$

$$= 6.5556\%$$

Appendix 2 Sensitivity Analysis

Freehold reversions

(i)   All risks yield of comparable rack rented freehold = 5%
         Discount rate (equated yield) = 11%
         Implied growth rate = 6.5556%

(a)   Change in discount rate = ± 0.5% (absolute change)

% Change in Capital Value

Length of Term

Ratio of Current Rent FRV	5	10	15	20
3:4	13.1634	13.9034	14.6289	15.2308
	10.4904	11.0331	11.5299	11.8996
2:3	13.3566	14.2369	15.0794	15.7854
	10.6403	11.2888	11.8685	12.3086
1:4	14.4369	16.2489	17.9864	19.5944
	11.4801	12.8224	14.0430	15.1060

Note: Row 1 – Value variation for a fall in discount rate
       Row 2 – Value variation for a rise in discount rate

The percentage change in value begins to decline after 35 years.

(ii) All risks yield of comparable rack rented freehold = 9%
      Discount rate = 14%
      Implied growth rate = 5.8774%

      Change in discount rate = ± 0.5% (absolute change)

                    % Change in Capital Value

                          Length of Term

Ratio of Current Rent FRV	5	10	15	20
3:4	6.9498 6.1342	7.3413 6.4492	7.5217 6.5666	7.4610 6.4738
2:3	7.1118 6.2741	7.6036 6.6710	7.8495 6.8382	7.8188 6.7646
1:4	8.1032 7.1284	9.4559 8.2359	10.4556 8.9977	11.0221 9.3674

Note: Row 1 - Value variation for a fall in discount rate
       Row 2 - Value variation for a rise in discount rate

Percentage change in value begins to decline after 20 years

41

(iii) Change in growth = ± 10% (percentage change)

(a)   All risks yield = 5%
             Discount rate   = 11%
             Implied growth rate = 6.5556%

% Change in Capital Value

Length of Term

Ratio of Current Rent FRV	5	10	15	20
3:4	16.6980	17.4586	18.0934	18.5275
	12.3933	12.8332	13.0977	13.1607
2:3	16.9721	17.9557	18.7864	19.3977
	12.5967	13.1994	13.6003	13.7787
1:4	18.4891	20.8991	23.2421	25.3443
	13.7227	15.4220	16.8260	18.0027

Note: Row 1 - Value variation for a rise in growth
      Row 2 - Value variation for a fall in growth

The percentage change in value begins to decline after 60 years

42

Leaseholds

All risks yield = 5%
Freehold discount rate = 11%
Implied growth rate = 6.5556%
Leasehold discount rate = 16%

Head rent fixed, subleased at reviews every 5 years

Change in discount rate = ± 10% (percentage change)

Length of Unexpired Term of Lease

Ratio of Head Rent FRV	5	10	15	20
2:3	7.6020	11.0273	14.0651	24.3987
	6.8270	9.4959	11.6612	17.4818
1:2	7.2410	10.3288	13.0534	22.1582
	6.9043	8.9314	10.8839	16.0696
1:4	6.9586	9.7288	12.1358	19.9239
	6.2779	8.4475	10.1755	14.6609

Note: Row 1 - Value variation for a fall in discount rate
          Row 2 - Value variation for a rise in discount rate

Leaseholds

Change in growth rate = ± 10% (percentage change)

Length of Unexpired Term of Lease

Ratio of Head Rent / FRV	5	10	15	20
2:3	3.0406	5.0109	6.4438	10.2988
	2.9669	4.8043	6.0823	9.1538
1:2	2.2238	3.8547	5.1083	8.5710
	2.1696	3.6961	4.8218	7.6181
1:4	1.5856	2.8638	3.8969	6.8477
	1.5467	2.7460	3.6783	6.0864

Note: Row 1 – Value Variation for a rise in growth rate
       Row 2 – Value Variation for a fall in growth rate

The tables are compiled on the assumption that there is no correlation between discount rates and growth rates.

44

Appendix 3

Two Growth Rate Model

Let y be the all risks yield of the freehold
r be the discount rate or equated yield
n rent review term
R current full rental value
g_1 expected rental growth rate during the first review period
g_2 implied annual growth rate thereafter assumed constant
$YP(n)$ PV of an annuity for n years (YP)

The capital value of the rack rented freehold may be expressed as

$$V = R.YP(n) + R\left[\frac{1+g_1}{1+r}\right]^n \cdot \frac{YP(n)}{1 - \left[\frac{1+g_2}{1+r}\right]^n} \qquad (A)$$

If g is the constant annual growth throughout

$$V = R. \frac{YP(n)}{1 - \left[\frac{1+g}{1+r}\right]^n} \qquad (B)$$

Let $\qquad G = \left[\frac{1+g}{1+r}\right]^n \qquad G_1 = \left[\frac{1+g_1}{1+r}\right]^n \qquad$ and $\qquad G_2 = \left[\frac{1+g_2}{1+r}\right]^n$

From (A) and (B), we have $\dfrac{R.YP(n)}{1-G_1} = R.YP(n) + \dfrac{R.G_1 YP(n)}{1-G_2}$

$$\frac{1}{1-G} = 1 + \frac{G_1}{1-G_2}$$

$$1 - G_2 = (1 - G)(1 - G_2) + G_1(1 - G)$$

$$= 1 - G_2 - G + GG_2 + G_1 - GG_1$$

$$G_2 = 1 + G_1 - \frac{G_1}{G}$$

i.e $\left[\dfrac{1+g_2}{1+r}\right]^n = 1 + \left[\dfrac{1+g_1}{1+r}\right]^n - \left[\dfrac{1+g_1}{1+g}\right]^n$

$$g_2 = \left[\left\{ 1 + \left[\frac{1+g_1}{1+r}\right]^n - \left[\frac{1+g_1}{1+g}\right]^n \right\} (1+r)^n \right]^{1/n} - 1$$

Appendix 4

Table A Variable growth rates

Freehold Reversion

(1)   All risks yield y = 5%
             Discount rate   r = 11%
             Implied annual growth rate g = 6.5556%

Short term growth rate g_1 Constant growth rate thereafter g_2
   10%   5.7092%
   12.5%   5%

Percentage Variation in Capital Values

Length of Term

Ratio Rent:FRV g_1	5	10	15	20
2:3 10.00	0.00	3.11	5.61	7.57
12.50	0.00	5.27	10.44	14.06
1:2 10.00	0.00	3.31	6.11	8.42
12.50	0.00	6.14	11.31	15.75
1:4 10.00	0.00	3.65	7.04	10.13
12.50	0.00	6.80	13.25	19.21

Table 3

(1)   All risks yield y = 9%
       Discount rate   r = 14%
       Implied annual growth rate g = 5.8774%

Short term growth rate g_1 Constant growth rate thereafter g_2
       10%   3.8053%
       12.5%   2.2864%

Percentage Variation in Capital Values

Ratio Rent:FRV	g_1	Length of Term			
		5	10	15	20
2:3	10.00	0.00	6.03	9.26	10.32
	12.50	0.00	10.59	17.28	
1:2	10.00	0.00	6.73	10.81	12.51
	12.50	0.00	11.89	18.88	21.23
1:4	10.00	0.00	8.17	14.47	18.37
	12.50	0.00	14.57	25.89	32.34

CHAPTER 4

DEPRECIATION AND PROPERTY INVESTMENT APPRAISAL

1   Introduction

Implicit and explicit appraisal models

The continuing shift away from implicit property valuation and analysis models towards explicit cash flow projections might not produce better valuations and analyses. That is a matter for debate. One benefit that has unarguably flowed, however, is the exposure of relevant variables for analysis.

The relationship between implicit and explicit models can be represented by a simple equation. Where y represents the all-risks yield (initial yield for a fully let property), r represents an overall return (redemption yield if a sale is envisaged) and g represents income growth, then (holding all else equal) the left hand side of the equation represents an all-risks yield approach while the right hand side of the equation represents a growth explicit model.

$$y = r - g \tag{1}$$

Irving Fisher:1930 helps us to expand the explicit model. The overall return on an investment (r) is a reward for three factors. These are l, a reward for losing the use of capital or liquidity preference, i, a reward for anticipated inflation, and p, a risk premium.

Hence, $$y = l + i + p - g \tag{2}$$

This model holds for all assets. It is an extremely useful representation of the variables that property analysts are now beginning to struggle with. However, given that property analysis is more commonly based on nominal returns, i is not typically exposed and the model becomes

$$y = (l + i) + p - g \tag{3}$$

where (l + i) is a risk free inflation prone opportunity cost rate (c) traditionally derived from the redemption yield on conventional gilts. (This is faulty, as the risk of inflation being other than as expected is ignored, but I will proceed to explore the popular approach.)

Hence, $$y = c + p - g \tag{4}$$

The importance of equation (4), faulty though it is, is that it exposes p and g as the two variables requiring exploration in the property context (and the subject of discussion in Chapters 5 and 6). While y and c are derived from market evidence, p, a property risk premium over gilts, requires theoretical and empirical work at the market, sector and individual property level, and g, estimated rental growth, is the most difficult variable to deal with, requiring both knowledge and faith, each greatly lacking in general at present.

Equation (4) is often presented as a deterministic model which should be capable of solution (see, for example, Crosby:1985 and Baum and Crosby:1988), most commonly by assuming a value for p (2%?) and 'calculating' a market expectation of growth (to seven decimal places in some cases!). But a value for p is not settled. There are arguments in favour of a nil or negative risk premium (see Fraser:1984) as well as support for a historically-based 2% (see Brown:1985).

Such is the pre-depreciation state of the art. But:

> 'There has been much discussion in the property investment market ... of the problems associated with obsolescence and its effect on property values ... There has been comparatively little thought given to the means of analysing its effect when undertaking valuations. There is a clear need for the profession to address this aspect of the problem...' (Debenham Tewson and Chinnocks:1985)

The work of Bowie (Bowie:1982), the CALUS report (Salway:1986a), the collective contributors to the Journal of Valuation 5:3 and many others recognise the importance of depreciation as a variable in property valuation/analysis models. Exactly how should it be accommodated as a third variable? This is considered in section 3 below.

Concepts of depreciation

Depreciation is traditionally referred to in basic US real estate texts while it is ignored in UK equivalents. The reason for this is clearly the importance of accelerated depreciation allowances against tax and the historically greater importance of tax in real estate returns in the US. Most texts, for example Wurtzebach and Miles:1984, refer to functional obsolescence, physical obsolescence and economic obsolescence as causes of depreciation. (Ironically, its US importance is waning through tax reforms as its UK importance appears to grow.)

Bowie:1982 introduced the topic to many UK property people by addressing the issue largely from an accounting viewpoint. SSAP 12 requires depreciation of all fixed assets; SSAP 19 makes exceptions for freehold investment properties. Bowie points out that there is an accounting anomaly because buildings depreciate even if they are in freehold ownership.

He then addresses what he sees as a second (property investment appraisal) anomaly and examines the impact upon initial yields of the 'hidden' depreciation factor. Thereby he criticises both accounting standards and valuation techniques. This was an early reference to a now topical subject, although Blandon and Ward (1979) had carried out a simple analysis of City office rents for depreciation in an earlier paper concentrating on growth expectations in property markets.

The major initial UK work concentrating on depreciation as a property (rather than an accounting) topic was the CALUS report, Depreciation of Commercial Property (Salway:1986a). This major research document was a wide-ranging study which included a consideration of property investment appraisal methods in the light of a theoretical consideration of the impact of depreciation and some empirical evidence towards assessing its impact.

Such is the current relevance of depreciation to property investment and appraisal methods that Journal of Valuation (5:3) included three depreciation-specific papers which each presented different appraisal solutions. These were Miles:1986, Salway:1986b and Harker:1986.

A more recent consideration of the effect of depreciation upon investment decision making and appraisal methods was hitherto unpublished work carried out at the City University (in 1987) for Richard Ellis and Hill Samuel. This was again the result of a major investment of resources. The appraisal methods produced as part of this work are considered in section 3.

Defintions

There is not a little confusion and lack of clarity regarding definitions used in the UK property market. The following basic definitions should be established before proceeding:

1. Depreciation is a loss in the real value of property.

2. Obsolescence is one of the causes of depreciation. It is a decline in utility not directly related to physical usage or the passage of time.

Sources of depreciation

From these definitions it is apparent that there are other causes of depreciation apart from obsolescence. Much of the City University work was addressed to the classification of these causes. Figure 1 begins to distinguish the contributory factors. Depreciation can result from tenure-specific factors (the shortening of a remaining lease length) or from property factors.

Property-specific depreciation can fall upon site or building (Figure 2). Sites may depreciate through environmental obsolescence factors (for example, detrimental changes in neighbouring buildings); and given that site values may be regarded as a residual, changing supply/demand relationships will also impact upon site values. Buildings, on the other hand, depreciate either through physical deterioration or building obsolescence (see Figure 3).

This latter factor is of particular philosophical interest. Salway produces a long list of obsolescence categories. In my view these are usefully confined to two main types: aesthetic obsolescence and functional obsolescence (see Figure 4).

Aesthetic obsolescence impacts upon the external appearance of a building and its internal specification (fittings). Functional obsolescence impacts upon its vertical and plan configuration (ceiling height; layout) and again upon internal specification (services, for example).

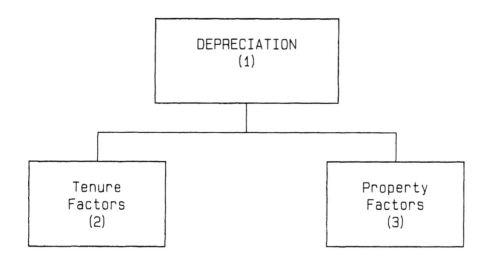

Figure 1   Sources of depreciation

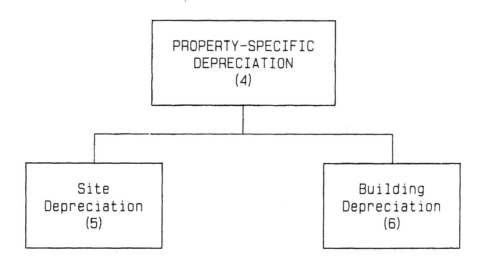

Figure 2   Sources of property-specific depreciation

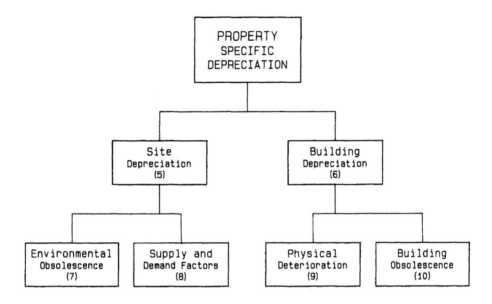

Figure 3 Site and building depreciation sources

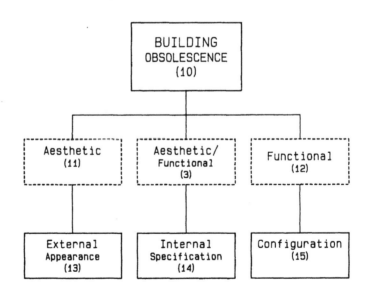

Figure 4 Sources of obsolescence

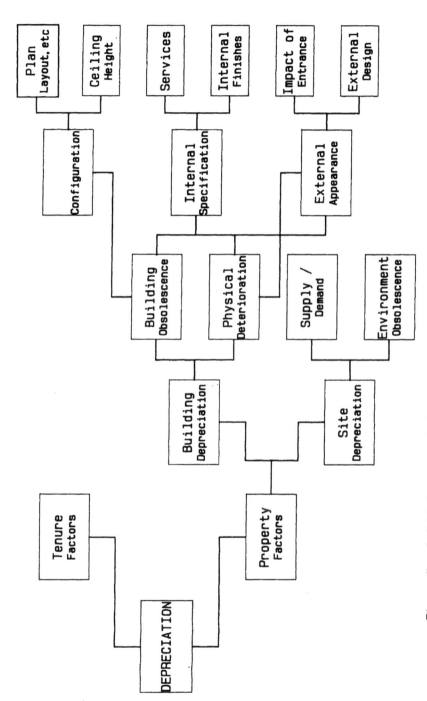

Figure 5   A full taxonomy of depreciation in existing use value

The classification of depreciation causes is not as simple as this suggests. Physical depreciation affects external appearance; aesthetic obsolescence also affects external appearance. Losses in real value of property resulting from poorer external appearance may therefore be the result of obsolescence or of deterioration, and a full classification is quite complex (see Figure 5).

The impact of depreciation

The value of an investment is determined by the return it produces. Return is a function of three main variables: income, return of capital and operating expenses.

Because the rental value of a property will be reduced by depreciation, income is affected. Depreciation is also evidenced by rising all-risks yields over time: as the rental value is reduced and the market yield increases, depreciation contributes to a double reduction in the available return of capital. Additionally, repair, maintenance and insurance costs will increase, so that depreciation contributes to rising operating expenses, thereby reducing net income.

The impact of depreciation on market value is thus threefold. Traditional valuation models are not capable of explicitly accounting for that fact, and rely wholly upon adjustment of the all risks (market) yield. In an efficient market which is not fully aware of causes or effects of depreciation, this is of no comfort to investors who require explicit appraisal models.

2 The Measurement of Depreciation

Empirical or theoretical models?

It has been established that depreciation may impact upon rental income, resale price and operating expenses, each affecting the value of a property investment. An attempt must be made to quantify each impact. Efforts are beginning to be made in this direction, and break into two methodologies. An all-risks approach based on hypothesis was used by Bowie:1982; explicit approaches using empirical evidence of depreciation rates have been used by Salway (Salway:1986a) and in the City University study.

A theoretical approach

Bowie's hypothesis is based upon the effective split of a freehold investment in property into an infinite-life land element and a finite-life (effectively leasehold) building element. By estimating a building content percentage, a building life and selecting an appropriate depreciation pattern (for example, straight line) it is possible to calculate the amount necessary to be set aside each year to recoup building costs after the end of the building life and estimate the impact on net income and running yield. Bowie estimated that an apparent 4.5% return on a prime 65-year life office building is reduced to a return of 3.9% after depreciation.

This paper was of considerable impact and arguably contributed to a continued slump in office and industrial property markets. However, Bowie's argument exposes the shortcomings of his approach.

'Only by breaking down the total investment in a property into its principal components and looking at their respective lives can a true yield be ascertained. If accurate comparisons are to be made of the performance of property investment as against gilts and equities, it is essential that depreciation be stripped out.' (Bowie:1982)

This is true: but what is a 'true yield'? In Bowie's paper it is by implicaton the running yield, which is an extremely poor measure of true yield and an equally bad basis for a comparison of property, gilts and equities, hiding as it does the crucial differences in growth potential, income reviews and so on.

Bowie's paper is also limited by its wholly theoretical or normative method of estimating depreciation. No evidence is presented for estimating building content, building life or depreciation pattern. It is suggested that, in order to progress towards a practicable appraisal model, empirical evidence of depreciation rates should be used in an explicit appraisal framework.

Explicit empirical models

In an explicit appraisal framework opportunities are presented for accounting for all sources of depreciation. Declining rental values can be accounted for; increasing resale yields have to be incorporated in what must be a finite holding period model; and rising expenditure can be dealt with as a deduction from year to year.

An example might appear as follows (Table 1). Assuming 2% per annum depreciation in rental value, yields increasing 1% per 10 years, 10% of capital value to be spent every 10 years to maintain this pattern and £1 per annum initial rental value, the following cash flow is presented ignoring growth and inflation. The price of the investment is £10.

Table 1   Depreciated cash flow

Year	Income (£)	Outlay (£)	Resale (£)	Expenses (£)	Net Cash (£)
0		(10)			(10)
1	1.00				1.00
2	1.00				1.00
3	1.00				1.00
4	1.00				1.00
5	1.00				1.00
6	0.91				0.91
7	0.91				0.91
8	0.91				0.91
9	0.91				0.91
10	0.91		7.45[1]	0.75	7.61

Note 1   Resale price = $\frac{£0.82}{0.11}$ = £7.45

These depreciation-related variables can be of significiant impact.
In the above example, an Internal Rate of Return (IRR) (ignoring
depreciation) of 10% is reduced to an IRR (with depreciation) of
7.28%. Sykes:1984 demonstrates this effect more fully. But the
vital question is: what values for the variables are reasonable?

Salway:1986a made a start in the pursuit of empirical evidence
for depreciation impact by using a cross-sectional method. He
questioned estate agents about rental values for hypothetical office
and industrial buildings at June 1985, comparing new with 5, 10 and
20 year old similar buildings. Depreciation rates were surprisingly
close to a straight line and were surprisingly similar for offices
and industrials (see Table 2).

Table 2   Rent depreciation rates (% of new rental value)

Age	Offices	Industrial
0	100	100
5	85	86
10	72	71
20	55	52

Source: Salway:1986a

This is extremely useful, and indicates depreciation rates of
around 3% per annum. However, it masks enormous regional variations
which are particularly related to differing site value proportions.
(For example, the City University work on City officers showed

average rates of around 0.75% per annum for new buidings.) It is
also based on hypothetical buildings: it is therefore debatable
whether this can be accepted as empirical evidence.

Additionally, Salway did not measure yield increases or rates of
expenditure on buildings in the same way. The picture is therefore
an incomplete one.

Nonetheless, there is no doubt that acceptable empirical evidence
is extremely hard to collect in this field. Two main methods present
themselves. A cross-sectional study, such as that used by Salway,
relies upon comparisons of identical buildings differing only in age.
Effectively, buildings of different ages are used as surrogates for
the subject building at different points in time. All other
variables - design, site value content, location quality and so on -
have to be controlled away. It is not surprising, therefore, that in
attempting a national picture based on 20/30 rent points Salway was
forced to use hypothetical buidings. Jones Lang Wootton:1987
improved upon this by using a cross-sectional study of actual
buildings, and found a rental obsolescence rate averaging 2.2% for
offices over the period 1980 to 1985.

A longitudinal study, on the other hand, is a direct comparison
of rental values of the same buildings as they age. This is more
promising, but is flawed if variables alter over time. This may
affect relative site values, for example. In any event, inflation
complicates the picture, so that depreciating rental values are not
immediately apparent. Finally, evidence is very hard to accumulate.

Both cross-sectional and longitudinal methods are, of course,
prone to the difficulties of most empirical research: that is,
relationships are not necessarily stable over time, and patterns of
depreciation witnessed now may not repeat over future holding
periods. A considerably greater problem rests with the CALUS study.
Given that obsolescence is one of the causes of depreciation (and) is
'a decline in utility not directly related to ... the passage of
time', it is insufficient to relate depreciation solely to age. It
is also a function of declining utility, which may be instantaneous.

The City University work attempted to overcome this difficulty by
combining cross-sectional and longitudinal methods with an
understanding of the causes of depreciation. Empirical evidence of
declining rental values and increasing yields was therefore compared
not only with age, but also with building quality, which was found to
be a better explanation of depreciation.

Expenditure

Expenditure on refurbishment, repairs and improvements is a
depreciation-related variable and should ideally be taken into
account in any explicit model. Sykes:1984, for example, hypothesises
the expenditure of a constant percentage of property value at regular
intervals to maintain rental value.

However, practical and theoretical difficulties abound. While
the cost of repair may be estimated with as much confidence as most
other variables, what of its impact?

More vitally is it realistic (as models such as Sykes' assume) to
hypothesise the renovation of a building to its exact state as new?

57

The answer to this is emphatically in the negative, as such an approach ignores the vital distinction between curable and incurable depreciation. The following arguments must be taken into account when considering allowances for future expenditure.

(1) In the City University study, it was found that incurable depreciation was marginally more important in relative terms than curable depreciation. No reasonable expenditure could reverse certain defects (for example, floor to ceiling height; external appearance, to a large extent; and plan layout). Thus a building must depreciate. Expenditure is but an imperfect means of redressing some depreciation: how much is cured depends on market circumstances.

(2) A considerable amount, perhaps the majority, of curable depreciation does not impact upon the building owner: it is dealt with by the tenant under the terms of the typical full repairing and insuring lease.

(3) The landlord may not have the opportunity to refurbish. If the tenant renews his lease under the 1954 Act, the landlord will not gain possession (although substantial redevelopment is a ground for possession). In any event, the holding period used in an appraisal model may not coincide with a lease break.

In order to reduce the error in the estimation of risky variables to a minimum, empirical evidence should be used to determine likely rental values and resale yields, assuming normal upkeep (and correction of curable depreciation) by the tenant, and a realistic amount of refurbishment expenditure by the landlord to deal with remaining curable depreciation only.

The City University research: the City office market

In the City University examination of depreciation in the central City of London office market, both cross-sectional and longitudinal methods of measuring depreciation were used to estimate depreciation in rental values, and a cross-sectional analysis was used to measure yield increases. Additionally, much effort was expended in gaining an understanding of causes of depreciation in order to build a model which was capable of practical interpretation rather than mechanical reproduction. The basis of this was described around Figures 1 to 5.

A sample of 125 buildings was selected for examination. Data for all 125 are used in a cross-sectional analysis, and an attempt was made to examine as many of the same buildings as possible in a longitudinal analysis. Data constraints, lack of continuinity of ownership being one, reduced the sample size to 33 for this longitudinal analysis. This may well be a representative reduction in sample size, and indicates a likely preference for cross-sectional methods in practice.

However, there are also problems in using cross-sectional analysis on a sample of real buildings, relating to lack of control over site factors and other influences upon rental value which may distort the direct comparison which lies at the heart of the analysis. Many of these were eradicated by using opinions of rental value rather than actual prices, set by the same panel of valuers at the same time. (Of course, the cost of this is a reduction in the value of the empirical evidence.)

The major remaining cause of a difference in rental value of different buildings is differences in site values/locational attractiveness. This was erased as much as possible by concentrating the sample within a tight area of the central City within which differences in site value were less than 10%, measured by estimating the rental value of one of the sample buildings on a selection of sites throughout the study area. The same sample data were then used to smooth away remaining value differences caused by location by the use of ratio adjustments.

Remaining rental values ranged from £20 to £40 per square foot. The hypothesis underlying the study was that any difference between £40 and rental value is created by depreciation. (This view of depreciation allows the possibility of a brand new building suffering immediate depreciation through, for example, bad design causing obsolescence: see definitions p50.)

Rental values were then related both to building age and to physical deterioration and a set of building obsolescence factors (based on Figure 4). For age, the relationship attained was much stronger for non-refurbished buildings than for refurbishments. The regression equation was:

Depreciation = £1.16 + 0.431 age
R^2 = 67.26% (significant at 95% level)

To test the importance of building qualities, shortfalls in rental value below a prime rent (depreciation) were related to shortfalls in each measure of quality (maximum 5). These qualities were configuration, internal specification and external appearance (obsolescence factors), and physical deterioration.

The regression equation was:

Depreciation = $-0.0654 + 0.218\ a^1 + 0.168\ a^2 + 0.22\ a^3 + 0.111\ a^4$

Where a^1 = 5 - configuration score
       a^2 = 5 - internal specification score
       a^3 = 5 - external appearance score
       a^4 = 5 - physical deterioration score
       R^2 = 84.1%

Physical deterioration is not significant at the 95% level, other factors are. Hence obsolescence (represented by the other three variables) is much more important than physical deterioration as a cause of depreciation.

Average rental depreciation for non-refurbished buildings was 0.4% pa over the first 18 years of their life, but this masked great variations (see Figure 6).

59

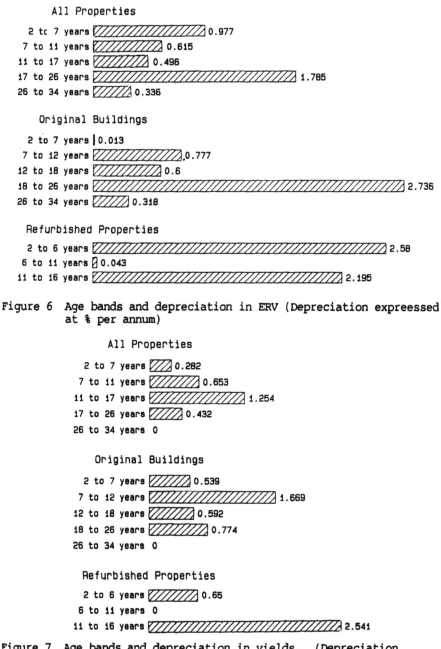

All Properties

2 to 7 years	0.977
7 to 11 years	0.615
11 to 17 years	0.496
17 to 26 years	1.785
26 to 34 years	0.336

Original Buildings

2 to 7 years	0.013
7 to 12 years	0.777
12 to 18 years	0.6
18 to 26 years	2.736
26 to 34 years	0.318

Refurbished Properties

2 to 6 years	2.58
6 to 11 years	0.043
11 to 16 years	2.195

Figure 6   Age bands and depreciation in ERV (Depreciation expreessed at % per annum)

All Properties

2 to 7 years	0.282
7 to 11 years	0.653
11 to 17 years	1.254
17 to 26 years	0.432
26 to 34 years	0

Original Buildings

2 to 7 years	0.539
7 to 12 years	1.669
12 to 18 years	0.592
18 to 26 years	0.774
26 to 34 years	0

Refurbished Properties

2 to 6 years	0.65
6 to 11 years	0
11 to 16 years	2.541

Figure 7   Age bands and depreciation in yields. (Depreciation expressed as % per annum)

A similar analysis was carried out for yields. Average yield depreciation for non-refurbished buildings was as shown in Figure 7. This averaged 0.8% pa (note that this is double the incidence of rental value depreciation).

Analyses of this type produced a full record of average depreciation rates in rental and yield, and revealed the impact of building quality decline. Good design was found to reduce the impact of depreciation by enabling landlords to hold a less risky asset and to benefit from tenants' efforts at repair and correction of curable depreciation: these factors should be taken into account in an operational model.

3 Depreciation and Appraisal Models

A third variable

On p.48 a simple equation is presented to represent implicit against explicit appraisal models. Concentrating on the use of the initial or all-risks yield (y) as a capitalisation rate, an implicit model is:

$$CV = \frac{rent}{y}$$

$$\text{where } y = c + p - g$$

and c = risk free inflation prone opportunity cost rate
   p = risk premium
   g = rental growth

Either p or g is typically treated as the dependent variable in a deterministic model. Most common is an assumption of p and a solution for g, the implied rental growth rate.

Given that depreciation has been shown to be a variable which impacts upon rental value and therefore rental growth, it is clear that g in this equation is net of depreciation rate. Alternatively, retaining a much more comfortable definition of g as expected average rental growth in continually new buildings g_m as measured by most indices (including the Investor's Chronicle/Hillier Parker index, it seems), less depreciation, the equation must be amended to:

$$k = c + p - (g_m - d)$$

$$\text{or} \qquad k = c + p - g_m + d$$

Three variables are now exposed for analysis: the risk premium p, rental growth in a sample of new buildings g_m, and average depreciation in the rental value of the subject property, d.

Note that implied growth is immediately greater than is typically recognised in analyses of this type: it is greater by a factor of d.

A basic model

Use of these variables in a basic appraisal model is complicated
by several factors. These include:

(i)   quarterly in advance rental income;
(ii)   non-annual rent reviews;
(iii)   changing yields over a holding period; and
(iv)   expenditure on refurbishments

A quarterly in advance adjustment can probably be assumed to be
of minor importance in illustrative models and incorporated at the
operational stage. Expenditure should be dealt with in a robust
manner based on empirical evidence and the lease details and not a
theoretical model. Changing yields have to be accommodated in a
finite holding period model (see p.55). This will naturally also
accommodate non-annual review patterns. The working model is shown
in an example below.

Salway's appraisal model

A similar method was presented by Salway in the CALUS study
(Salway: 1986a). However, Salway's method requires the valuer to
estimate the value at reversion as the higher (or a weighted average)
or site value, residual value for refurbishment, or investment value
as re-let unimproved.
The former two reversion values have to be estimated in
present-day terms and inflated. The latter is estimated by
capitalising a depreciated, inflated rental value at a resale point.
This model is probably over-complex. The estimation of residual
value for refurbishment is extremely difficult even in current day
terms, and should not be used as an input. Instead, the model should
be simplified to take an automatic reversion to the higher of
inflated site value and investment value as re-let unimproved.   Such
a format is illustrated at p.64.

An appraisal model in practice

(This is a simplified version of a model developed as part of the
City University work.)

Inputting depreciation

Estimates of both rental and yield expenditure must be made in a
depreciation-explicit appraisal model.
Increasing yields are easily accommodated, but various
alternatives present themselves for the inputing of rental depreciation as a variable. These include a rate per cent depreciation
per annum over each review, and the estimated rental value per square
foot at each review. Use of a cross-sectional analysis to research
rental values over time suggests a probable market preference for the
latter, which is therefore used in the model described below.

The effect of upward-only rent reviews

The preponderant use in the UK prime commercial property market
of 20 or 25 year leases with 5-yearly reviews to current rental value
or the current rent, whichever is higher, has important implications
for an appraisal model which is explicit regarding depreciation. It
is a factor disregarded in publications to date that this system
provides an inbuilt safety net or floor to the amount of depreciation
that can be suffered by an investor, and in a period of low inflation
this becomes an important consideration in a purchase. An appraisal
model should not therefore allow projected rental income to fall
below the previous level of rent. Modifications to the model will,
however, be necessary where lease ends occur during the holding
period, thereby removing the support of the upward-only rent review.
    In the models demonstrated below, it is assumed that the lease
end coincides with the end of the holding period, so that the resale
price is projected on the basis of current rental value at the
reversion discounted at the current capitalisation rate, without an
inbuilt floor to the rent that may be achieved.

Real or nominal?

    In Section 1 above it was stated that explicit cash flow
projections are becoming increasingly accepted as the basis of
decision making in property investment. It is also true to say that
all models developed in the reference material are nominal. In other
words, they project expected cash flows and calculate rates of return
on this basis. Conventional (fixed interest) gilts are usually the
basis of a target rate in such models.
    However, the impact of inflation post 1960 has resulted both in
demand for estimates of real (net of inflation) return and in the
possibility of using the yield (expected real redemption yield) on
index linked gilts as the basis for a real target rate. A lack of
development in property analysis on a real basis is an omission which
should be rectified in order to bring the property investment market
into line with alternative markets.

The model is illustrated in nominal terms only, for simplicity.

The input variables

    In the following, for simplicity, a standard 15 year holding
period is used. The model is specific for fully-let (non-
reversionary) investments and 5-yearly reviews.

Data to be input into the model are as follows:

1.   Price of investment in £.
2.   Current estimated rental value/contract rent (these should
     equate) in £.
3.   Site value in £ (estimated by comparison or, where
     necessary, as a percentage of price or value).
4.   The target rate, possibly derived from an adjusted average

of redemption yields on 15-year conventional gilts, input as a decimal.

5. <u>Rental growth</u> at <u>review 1</u>, <u>review 2</u> and <u>review 3</u> is forecast percentage rates of growth in prime property rental value over the first, second and third review periods respectively, input as decimals.

6. <u>Estimated rental value</u> at <u>review 1</u>, <u>review 2</u> and <u>review 3</u> is average depreciation rental values at each review date in current terms, in £.

7. <u>Estimated resale capitalisation</u> rate for a typical similar building albeit 15 years older.

8. <u>Inflation</u> over <u>review 1</u>, <u>review 2</u> and <u>review 3</u>, input as percentages per annum.

Calculations

Calculations automatically carried out within the model are as follows.

1. Estimates of current rental value for age 5, 10 and 15 are used to calculate declining balance depreciation rates over each review. Average rental depreciation over the holding period on a straight line basis is calculated and compared with total depreciation per annum, again on a straight line basis, this time incorporating the effect of yield changes.

2. Projected estimate rental value is the actual rental income expected, combining estimates of rental growth and depreciation.

3. Projected realisation is the higher of the projected site value and the projected property value (rental at year 15 divided by resale capitalisation rate) in 15 years' time. Demolition costs are ignored in this comparison.

Analyses

Analyses provided by the models are <u>net present value</u> (NPV) and <u>internal rate of return</u> (IRR). A positive NPV and an IRR in excess of the target rate indicate a decision to buy. The <u>cash flow</u> <u>statement</u> projects the actual income and the real income expected.

The appraisal model illustrated: an example

Price: £800
<u>Estimated rental value</u>: £40
<u>Site value</u>: £500
<u>Target rate</u>: 11%
<u>Rental growth</u> at <u>review 1</u>, <u>review 2</u> and <u>review 3</u>: 8%, 8% and 8%
<u>Estimated rental value</u> at <u>review 1</u>, <u>review 2</u> and <u>review 3</u>:
        £38, £36 and £35 respectively (net of depreciation)
<u>Resale capitalisation rate</u>: 5.75%
<u>Expenditure forecast</u>: nil

These data are based on evidence of the central City office
market at mid-1986 and a new property let on a 25-year lease and held
for 15 years. Below is a printout of the analysis of the property
based on the information listed. A summary of results is shown in
Table 3. The decision is marginal. Table 4 (presented after the
analysis) shows a cash flow statement: compare with the basic cash
flow model shown in Table 1.

Table 3   Appraisal model: results

NPV	IRR	Buy?
(£25)	10.69%	No

APPRAISAL MODEL

5 YEAR REVIEW/NON REVERSIONARY

1   Data

Price	£800.00
ERV	£ 40.00
Site value	£500.00
Target rate	11.00%
Rental growth review 1 pa	8.00%
Rental growth review 2 pa	8.00%
Rental growth review 3 pa	8.00%
Current ERV review 1	£ 38.00
Current ERV review 2	£ 36.00
Current ERV review 3	£ 35.00
Resale cap'n rate	5.75%
Expenditure at year –	0
Expenditure at year –	0
Expenditure at year –	0

2 DEPRECIATION

Current ERV at review 1:	£ 38.00
Current ERV at review 2:	£ 36.00
Current ERV at review 3:	£ 35.00
Resale cap'n rate	5.75%

3 GROWTH

Review	1	2	3
Prime growth pa	8.00%	8.00%	8.00%

4 PROJECTIONS

Proj ERV dep'n pa, review 1	0.98%
Proj ERV dep'n pa, review 2	1.03%
Proj ERV dep'n pa, review 3	0.55%
Proj ave straight line ERV dep'n pa	0.79%
Projected ERV, review 1	£ 56.00
Projected ERV, review 2	£ 78.00
Projected ERV, review 3	£111.00
Current cap rate	5.00%
Projected resale cap rate	5.75%
Projected site value	£1,576.00
Projected net resale value	£1,931.00
Projected realisation	£1,931.00

5 ANALYSIS

Target rate	11.00%
NPV	(£25.00)
IRR	10.69%

Table 4   Cash flow statement

Year	Income(£)	Outlay (£)	Resale (£)	Net Cash (£)
0	40	(800)		(800)
1	40			40
2	40			40
3	40			40
4	40			40
5	40			40
6	56			56
7	56			56
8	56			56
9	56			56
10	56			56
11	78			78
12	78			78
13	78			78
14	78			78
15	78	1,931		2,009

4   Other issues and conclusions

Depreciation and risk

The appraisal model presented in this paper is apparently
predicated on the basis that depreciation rates in both rents and
yields can be accurately forecast. This is, of course, nonsense, and
the introduction of a depreciation variable at the same time
introduces a risk that the forecast is incorrect. Hence some type of
probabilistic model, such as a simulation programme, should be
employed, or if deterministic risk adjustment techniques are favoured
a higher risk premium may be justified for depreciating properties in
addition to the value reduction created by an explicit allowance for
depreciation in rental values and yields.

Implications for performance measurement

Accumulated depreciation may be disguised by the annual valuation
and the artificiality of the rent review as a test of market rental
values. Given the general growth in the depreciation rate over the
holding period of a property investment, the effect of this disguise
while a building is tenanted is to ensure a rapid drop in value at
the lease end. Should, therefore, performance measurement techniques
be amended to allow for accumulated depreciation?

Conclusions

The implications of depreciation in property investment are yet another argument for explicit performance measurement and appraisal approaches. Those who take comfort in reducing inflation rates as a cause for renewed reliance upon pre-inflation appraisal methodologies have no cause for confidence. The City University study confirmed by Jones Lang Wootton:1987 showed that depreciation rates are probably getting faster. It is a true third variable, it is here to stay, and it should be dealt with explicitly.

References

Baum A E and Crosby F N:1988, <u>Property Investment Appraisal</u>, London: Routledge

Blandon P R and Ward C W R:1979, 'Expectations in the property market', <u>The Investment Analyst</u>, 52:24

Bowie N:1982, 'Depreciation: who hoodwinked whom?' <u>Estates Gazette</u>, 262:405

Brown G R:1985, 'An empirical analysis of risk and return in the UK commercial property market', unpublished PhD thesis, University of Reading

Crosby F N:1985, 'The application of equated yield and real value approaches to the market valuation of commercial property investments', unpublished PhD thesis, University of Reading

Debenham, Tewson and Chinnocks:1985, <u>Obsolescence: its Effect on the Valuation of Property Investments</u>, London: Debenham, Tewson and Chinnocks

Fisher I:1930, <u>The Theory of Interest</u> Philadelphia: Porcupine Press

Fraser W D:1984, <u>Principles of Property Investment and Pricing</u>, London: Macmillan

Harker N:1986, 'The valuation of modern warehouses: inflation and depreciation implications', <u>Journal of Valuation</u>, 5:138

Jones Lang Wootton:1987, '<u>Obsolescence: The Financial Impact on Property Performance</u>, London, JLW

Miles J:1986, 'Depreciation and valuation accuracy', <u>Journal of Valuation</u>, 5:125

Salway F W:1986a, <u>Depreciation of Commercial Property</u>, Reading: College of Estate Management

Salway F W:1986b, 'Building depreciation and property appraisal techniques', <u>Journal of Valuation</u>, 5:118

Sykes S:1984, 'Periodic refurbishment and rental value growth', <u>Journal of Valuation</u>, 3:32

Wurtzebach C H and Miles M E:1984, <u>Modern Real Estate</u> (2e), New York: Wiley

CHAPTER 5

THE ANALYSIS OF RISK IN THE APPRAISAL OF PROPERTY INVESTMENTS

1   Introduction

Property investment, like all investment, involves an initial capital
outlay in return for the expectation of future income receipts.
Depending on the type of investment, these receipts may range from
being totally certain (fixed income investments) to being totally
uncertain, although this extreme is unlikely as normally there will
be a lease with a contractual obligation to pay rent. In this case
the uncertainty will be concerned with the tenant's covenant (i.e.
the likelihood that this contractual obligation will be complied
with) and the degree of rental uplift, if any, at subsequent rent
reviews. Furthermore there may be uncertainty over obsolescence,
declining rental values and the need for further capital injection as
well as uncertainty as to how long the investment will be retained
and what it might be worth when eventually sold. Traditionally,
these uncertainties and other factors such as liquidity, cost and
inconvenience of management etc., have all been reflected in the
overall yield used to value the income receivable – hence the
expression the all risks yield (ARY). The yield is all embracing
taking into account the risks of investment.
   The purpose of this chapter is to examine the risks involved in
property investment and to move away from the concept of applying an
all embracing yield towards a concept of quantifying risk and
allowing for it separately through the use of sensitivity and
probability analysis, methods which have long been used in the
analysis of other non-property investments. Indeed it has been said
that the 'real estate literature is surprisingly devoid of
theoretical and practical treatment of the assessment of risky
investments and, indeed, trails the general finance literature, by
perhaps 25 years' (Young:1977).   It is also probably true that in
this country property literature is 10 years behind US real estate
literature. As with all DCF techniques generally the aim of risk
analysis is to be explicit rather than implicit. 'Real estate
decision makers claim they take "calculated risks", but few of them
make very clear just how they calculate these risks. Traditionally
because of the difficulties, the dislike, or the lack of knowledge of
how to deal explicitly with risk in decisions most people
concentrated on a few key assumptions about the future, examined a
few rules of thumb, mulled over the situation, and then decided.
Although some of the risk considerations were explicit, most of the
mathematics of risk was left to the four horsemen of the implicit
decision making apparatus: judgement, hunch, instinct and intuition'
(Phyrr:1973). As Stephen Sykes said recently:

   'Maths cannot replace intuition, but modern techniques can aid in
   quantifying information which is normally considered
   qualitatively' (Sykes:1983a).

Ir discussing available techniques this chapter will hopefully bear in mind at all times that techniques must be easily understood and relatively easy to apply in practice otherwise they will be dismissed and gather dust on practitioners' shelves.

The structure of this chapter falls into four parts. The first part examines briefly the variables involved in property investment and the different types of property investment. It poses the question whether some investments are inherently more risky than others e.g. a short leasehold retail investment compared to a reversionary freehold office investment.

The second part examines, also fairly briefly, the market approach to risk analysis which appears to be essentially an intuitive approach.

The third part, with the aid of a case study, looks at various approaches that have been put forward to quantify risk and provide more information to aid the investor in decision making. This part falls into three sections (a) sensitivity analysis and scenarios, (b) risk free and risk adjusted discount rate approaches, and (c) probability analysis such as the Hillier approach and Monte Carlo simulation.

The final part will attempt a conclusion as to the applicability of the techniques discussed.

2. Aspects of Risk in Property Investment

Although the number of variables in an investment appraisal is fewer than in a development appraisal and the effect of changes in these variables is less significant than in a development appraisal this should not be an excuse to ignore them or to avoid quantifying the effect that changes in those variables will have on investment value.

In a typical investment appraisal the following variables could be identified.

Rental value and rental growth

If the investment is reversionary there will be uncertainty about estimated rental values (ERV) due to insufficient comparable evidence. For some investments there will be more and better comparable evidence than for other investments and this will effect the level of risk involved. The future change in ERV at each rent review date (i.e. rental growth) will also be uncertain. Past evidence of growth rates may be a guide and techniques of forecasting rental growth will be useful in assessing at least a range of probable values, but inevitably there will be uncertainty and the further into the future the greater the uncertainty. Future income may differ from expectations due to changes in economic growth and inflation, competition from new locations (e.g. out-of-town -retailing) and timing of rent reviews and the cyclical nature of the market. For certain investments which are highly geared (e.g. some leaseholds) this uncertainty may be more relevant than for others such as high yielding industrial investments where little growth in income is anticipated. Similarly in this respect low yielding retail investments may be considered more risky than high yielding industrial investments simply due to the higher rental growth expected.

Yield on sale and timing of sale

Although property has traditionally been considered a long term investment it is likely that it will be sold at some stage and not retained in a portfolio indefinitely. Increasingly this holding period has become shorter and performance over the short term has become vitally important. Recent figures from the Central Statistical Office show that overall the rate of disposals from institutional property portfolios is now at an all time high and at times virtually equal to new purchases. Many institutions have recently been net disinvestors of property.

Considerable uncertainty therefore exists as to when an asset will be sold and what price will be achieved. The price in turn will be affected by the rental income and rental value (see above) and the capitalisation yield which may be lower or higher than at the appraisal date. Yields may alter as the overall pattern of yields moves up or down or because of a change within the structure of property yields (e.g. one sector moves in or out of favour) or because of the size of the investment and the strength of the property investment market (e.g. yield premia on large investments). Further problems surround the imprecise nature of the property investment market. Unlike the stock market there is no central market with prices being quoted on a regular basis. This adds to uncertainty as well as to illiquidity. As recent studies by Hager and Lord:1985 and Miles:1987 show, property valuation is not a precise art and the more 'unusual' the investment is (in terms of good comparable evidence) the wider the range in valuations there is likely to be from different valuers. The possibility of a 'special purchaser' paying over the odds further adds to uncertainty.

Age and obsolescence

This is the subject of Chapter 4 and therefore will only be mentioned briefly. Obsolescence has only very recently been researched in any detail and much work still needs to be done, but recent studies (see Salway:1986 and Jones Lang Wootton:1988) show the problem of obsolescence to be much greater than was generally considered only a few years ago. Uncertainty from an investment appraisal point of view therefore exists as to the magnitude and timing of future capital expenditure – whether this is minor or major refurbishment or complete redevelopment. The relationship between site value and investment value will obviously be important here. So that for a prime retail investment where site value will be a high proportion of investment value and where 'refurbishment' will probably be undertaken periodically by the tenant, uncertainty will be greatly reduced compared to a suburban/provincial office investment, for example.

Lease structure

The timing of rent reviews, duration of the lease(s) and the tenant's covenant will have a significant bearing on uncertainty. Where a property is reversionary the duration of the lease will affect uncertainty as the income stream will not be guaranteed once the lease ends. The tenant may not want a new lease and if the letting market is weak, there may be difficulty in reletting. Not only will there be uncertainty about the level of rent but also about the length of the period of time until that rent will be received. There may be an extensive void period. When eventually relet there will be uncertainty about a future tenant's covenant and therefore the security of that income, relating both to the length of lease granted and the reliability of the tenant to pay the contractually agreed rent.

The building's age and location will obviously be important here. But even for a new prime investment, where the above problems are less relevant, the time from the appraisal date to the date of the next rent review will affect capital value and its certainty. Clearly the nearer the rent review date and the more reversionary the investment, the greater the uncertainty of investment value. Just after a rent review uncertainty is minimised, just before a rent review uncertainty is maximised (Sykes:1983b).

Liquidity, management costs, taxation and inflation

These terms are hopefully self-explanatory and the latter two in particular relate to all investments. As mentioned earlier uncertainty will exist about the value of an investment at such time as a disposal occurs. This will be compounded by the ease of selling – how long it takes and how much it costs to sell. In this sense a prime investment may be considered less risky than a secondary investment as it will be more marketable, particularly in a weak property investment market. Earlier it was mentioned that a low yielding retail investment may be considered more risky than a high yielding industrial investment for example due to the higher rental growth needed to provide the same equated yield. In this sense the equated yield of the industrial investment is more secure as it may be achieved largely from the initial yield, requiring little or no rental growth. However, in terms of marketability (and obsolescence) the uncertainty of the industrial investment may be much greater than the retail investment.

From the foregoing it is clear that some investments will be more affected by uncertainty than others and hence risk analysis will be more relevant to some investments than others. Table 1 below categorises investments and lists characteristics relevant to a study of risk.

73

Table 1

Tenure of Investment

Freeholds – rack rented
           – reversionary – extent of uplist
                          – time of reversion
                          – lease end or rent review
                          – marriage value

Leasehold – as above for freeholds
          – medium and long leaseholds – extent of gearing and basis
                                          of rent review (upward only
                                          or side by side)

                                       – refurbishment clauses

          – short leaseholds   – relationship between rent
                                          paid and rent received,
                                          frequency and nature of
                                          rent reviews

                                       – fixed income

          – importance of marriage value

Type of Investment

Shop   ) – prime or secondary
Office,   ) – investment size
Industrial) – age
etc.   )

---

3 The Market Approach to Risk Analysis

It has long been accepted that risk is a major factor in property
development due to the large number of variables and the extremely
sensitive nature of the residual method of valuation.   Investment in
an already built and let property is clearly much less risky as the
number of key variables is fewer and the method of appraisal much
less sensitive to changes in these variables. Even in development
appraisal the use of risk analysis is in its infancy and in practice
intuition still plays an important and dominant role. Nevertheless,
many of the techniques that are now available on standard 'off the
shelf' commercially available programmes utilise techniques in an
easy to use way. Gradually these are being incorporated into
development appraisal. For example, the programmes marketed by
software firms contain easy to use sensitivity and probability
analysis and, on occasion, include a (somewhat simplified) Monte
Carlo simulation programme, which is apparently used in practice by

some developers and their clients. The same level of sophistication is not yet apparent in the investment end of the market although the basic techniques are the same or very similar.

In practice the single point estimate is still widely used with the traditional ARY approach. Intuitively the discount rate is adjusted to reflect risk along with all other relevant factors. The market approach to risk analysis is, in many cases, to ignore it.

At most where an investment is perceived to be subject to greater risk, e.g. a (highly) reversionary investment, a simple form of sensitivity analysis might be adopted - 'a what if' approach. A range of ERV's and yields might be assumed to calculate the reversion, so giving a range of capital values enabling the internal rate of return (IRR) to be calculated. no explicit probability analysis would be undertaken but intuitive judgement would be used to concentrate on the central figures of the resulting range - i.e. an intuitive form of probability analysis.

4   Risk Analysis and Discounted Cash Flow (DCF) Appraisal

Case Study

A case study of a reversionary freehold shop investment is used to illustrate the techniques that will be discussed. This shop is located in a good position in an attractive town in the south east of the country and at the date of acquisition in 1985 was let for £15,000 pa on full repairing and insuring terms (FRI) whilst the ERV was estimated to be £24,000 pa. The lease is reviewable at five yearly intervals, the first review being due two years after acquisition. This review has now taken place and £30,000 pa was achieved, a considerable uplift over the ERV two years previously but the ARY has also altered - partially counteracting the uplift in rent. Although a valuation is given for 1987 when the investment is no longer reversionary this is intended mainly for illustrative purpose and the appraisal date for the purpose of most of this paper is assumed to be the date of acquisition in 1985.

Conventional Value in 1985

Income	£15,000 pa	
YP in perp. @ 4.35%	22.99	£344,828
Rental value	£24,000 pa	
Less rent passing	£15,000 pa	
Increase on review	£ 9,000 pa	
YP in perp. def. 2 years @ 4.35%	21.11	£190,006
Total		£534,834
Net of 2.75% acq. costs		£520,519
say		£520,000

Conventional Valuation in 1987

Income	£30,000 pa	
YP in perp. @ 4.75%	21.05	£631,579
Net of 2.75% acq. costs		£614,675
say		£615,000

Simplified DCF Appraisal 1985

A crude analysis of the equivalent yield of 4.35% shows a market implied growth rate of 8.25% pa assuming a discount rate of say 12% (i.e. long dated gilts plus 2% premium).

Income	£15,000 pa	
YP 2 years @ 12%	1.69	£ 25,350
ERV	£24,000 pa	
Growth @ 8.25% 2 yrs	1.171806	
ERV in 1987	£28,123 pa	
say	£28,000	
YP perp. @ 4.35%		
PV £1 2 yrs @ 12%	18.3275	£513,170
		£538,520
Net of 2.75% acq. costs		£524,107
say		£524,000

This simplified DCF approach, whilst giving an indication of the growth in rent that is implied by the purchase price of £520,000, is nevertheless oversimplified for the following reasons:

(i) The equivalent yield has been assumed to be identical to the ARY. In practice this means that the greater security of the term income is exactly offset by the uncertainty over the reversionary income. In view of the attractiveness of the investment to institutional purchasers this is a reasonably accurate interpretation of the investment climate in 1985 particularly as the ERV of £24,000 was, if anything, conservative.

(ii) Constant rental growth has been assumed both in the short term,
up to the review date, and in the longer term, as reflected in
the ARY used which implies 8.25% pa growth in perpetuity. This
approach also assumes a disposal after the review in year 2,
although as the following appraisal shows, disposal in year 12,
for example, will give a similar capital value providing constant
rental growth is assumed. But in this example purchasers'
acquisition costs, incurred in year 12, have been allowed for.

Some of these oversimplified and restrictive assumptions will be
altered in the methods of risk analysis discussed later in this
section, but the following appraisal is used as the basis for
illustrating these methods.

Income	£15,000 pa	
YP 2 yrs @ 12%	1.69	£ 25,350
ERV in 2 yrs	£28,000 pa	
YP 5 yrs @ 12%		
PV 2 yrs @ 12%	2.8739	£ 80,469
ERV	£24,000 pa	
Growth @ 8.25% 7 yrs	1.742785 pa	
ERV in 7 yrs	£41,800 pa	
YP 5 yrs @ 12%		
PV 7 yrs @ 12%	1.63072	£ 68,164
ERV	£24,000 pa	
Growth @ 8.25% 12 yrs	2.589	
ERV in 12 yrs	£62,136 pa	
say	£62,100 pa	
YP perp. @ 4.35%	23	
	£1,427,585	
Net of 2.75% acq. costs	£1,389,377	
PV 12 yrs @ 12%	0.256675	£356,618
		£530,601
Net of 2.75% acq. costs		£516,400
Say		£516,000

Sensitivity analysis and scenarios

Sensitivity Analysis

There are many forms that sensitivity analysis can take, all of
them examine the degree of capital value change caused by a change in
one or more of the variables. Sensitivity coefficients can then be
determined showing capital sensitivity due to change in each variable
in isolation (see Sykes:1983b) and these capital sensitivity factors
can be compared between investments as a crude measure of risk. So
one form of sensitivity analysis would be to take say a 10% change in
one variable and calculate the percentage change in capital value.
In this example a +10% change in ERV results in a +9.1% change in

capital value, whilst a +10% change in rental growth up to the date of disposal results in a +7.2% change in capital value and a +10% change in yield at the date of disposal results in a −5.7% change in capital value. This shows that for these variables the valuation is more sensitive to change in rental value, suggesting that most attention should be paid to estimating this variable.

An alternative approach would be to take a range of possible values for each variable and combine them together to give a range of capital values (the range of assumptions is shown in Table 2 below). This would then show the extent that the valuation could vary, albeit on a crude basis as the probability of any of the estimated values occurring has not been determined. We simply have a spread of values giving the expected value and the highest and lowest values. Nevertheless, this is an advance over the original single point valuation and suggests the value will be £514,000 (mean) with a range of plus 32.5% to minus 25.5%.

Table 2 Assumption for sensitivity analysis

ERV	Rental Growth To Year 12 %pa	Yield on Disposal %pa
£22,000	6.5	4
£24,000	8.25	4.35
£26,000	10.0	5

Scenarios

The use of scenarios is a simple extension of the sensitivity analysis illustrated above (see Table 3 below). Rather than having a range of values for each variable giving a wide range of capital values, estimates are grouped to give say optimistic, realistic and pessimistic scenarios of capital value as shown below. This is likely to give a narrower band of values than the simplistic sensitivity analysis illustrated above due to subjective probability being employed to limit the likely combinations of variables and hence spread of results. Nevertheless, although this may be a slight improvement the probability of the estimates being achieved has not been assessed and as a result there is no explicit probability of the three capital values being achieved nor the standard deviation of the capital value. Are all three equally likely or is the realistic scenario more likely than the two extremes?

Table 3(a) Assumptions for scenarios

	ERV £	Rental Growth %pa	ARY When Sold %pa
Optimistic	25,000	9	4.25
Realistic (a)	24,000	8.25	4.35
Realistic (b)	24,000	8.5	4.5
Pessimistic	23,000	7.5	4.5

Table 3(b) Results

	Capital Value £	Purchase Price £516,000	
		IRR %	Rental Growth %pa
Optimistic	578,000	13.3	7.7
Realistic (a)	516,000	12.0	8.25
Realistic (b)	516,000	12.0	8.5
Pessimistic	459,000	10.7	9.2

Risk Free and Risk Adjusted Discount Rate Approaches

Risk adjusted discount rate (RADR)

This approach is an extension of the traditional ARY approach although here the growth element of the income is made explicit and the adjustments that are made to the discount rate only reflect risk. So high risk cash flows require a high risk premium on the basic discount rate and low risk cash flows require a low risk premium. For example, a short leasehold investment with fixed rent payable, but reviewable rent receivable, producing a volatile highly geared profit rent, would be considered to have a very risky future cash flow and so a high risk premium. Conversely, a rack rented high yielding freehold investment with little growth potential might be considered to have a low risk future income cash flow and a low risk premium.

There are a number of problems with this approach. What range of risk premia should we consider, +2%, +4%, +6%, or +8%?

As the following example illustrates even a slight adjustment to the discount rate for risk can have a dramatic effect. Compare this result with the range of values obtained from the sensitivity analysis. One of the main reasons for this is that the risk premium will be compounding the further into the future we go. This means not only that future cash flows are considered more risky than present ones (a realistic assumption in most cases) but that this degree of risk is compounded and so increases exponentially each year. A simple example will illustrate this (see Table 4). Assume the risk premium is only 2% then after ten years the cash flow will be considered to be approximately $(1+0.02)^{10}$ times as risky as the income receivable today, i.e. approximately 22% more risky. Perhaps even more significantly the income in say year 6 will be considered automatically more risk, than the income receivable in year 10, even though with 5 yearly reviewable leases the income will remain unchanged.

Table 4 RADR – assumption and results

ERV	£24,000 pa
Rental Growth 0-2 years	10% pa (short term optimism)
2-12 years	8% pa
Yield on Disposal	4.5%
Capital Value @ 12%	£515,000
@ 13%	£471,000
@ 14%	£432,000

Risk free discount rate (RFDR)

The RFDR, or certainty equivalent method, avoids this problem by adjusting the cash flows to reflect risk and discounting at the risk free discount rate. This method has the advantage of greater flexibility although it is not perhaps so easy to use in practice for valuers who are more familiar with adjusting discount rates or yields to reflect risk. An example of this approach is shown below. Baum:1987 suggests an objective way of determining the certainty equivalent cash flow which he defines as a cash flow where there is approximately an 84% chance of bettering and only a 16% chance of failing to achieve. To undertake this calculation a number of assumptions have to be made, some of which are either dubious or oversimplified (as he admits). The first, perhaps not unreasonable assumption, is that the cash flows are normally distributed so that one standard deviation either side of the expected cash flow includes 68% of all possible outcomes (see Table 5 below). Therefore if a value for the cash flow is taken which is the expected, or best estimate (or mean), less one standard deviation (SD), it follows that only 16% of possible outcomes lie below this figure i.e. 0.5 x (100% - 68%) and this can be taken as an estimate of a reasonably certain cash flow (see Table 5 below). Arguably to be even more certain of this cash flow the best estimate less say 2 SD's could be taken (only 2.5% of all outcomes would then be lower than this figure). The valuer would then have to assess a value for each variable which was certain to be achieved. This could therefore be assumed to be say 3 SD's from the best estimate. Three SD's either side of the cash flow includes 99.74% of all possible outcomes in a

normal distribution (see Table 5 below). So in the example used
previcusly, the following values might be taken.

Table 5 RFDR - assumptions and results

Expected ERV	= £24,000
Lowest likely ERV	= £21,000
Therefore SD of ERV	= £ 1,000 (i.e. 3000/3)
Therefore CE of ERV (a)	= £23,000 (1 SD)
(b)	= £22,000 (2 SD)
Expected rental growth	= 8.5% pa
Lowest likely rental growth	= 4.5% pa
Therefore SD of rental growth	= 1.33% pa
Therefore CE of rental growth (a)	= 7.17% pa (1 SD)
(b)	= 6.08% pa (2 SD)
Expected yield on sale	= 4.5%
Highest likely yield	= 5%
Therefore SD of yield	= 0.1667%
Therefore CE of yield (a)	= 4.67% (1 SD)
(b)	= 4.83% (2 SD)
Discount rate	= 10% (gilts with no margin)
Capital Value (a)	= £520,000
(b)	= £370,000

An investor's risk return indifference is difficult to assess and the
choice of certainty equivalent need not be expected value less 1 SD
but may be as low as expected value less say 2 SD depending on the
investor's risk aversion.

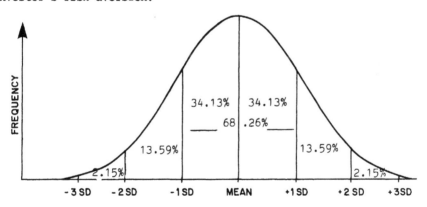

The areas between intervals of the standard normal distribution

81

Sliced income approach

Baum:1987 suggests a combination of the RADR and RFDR approaches
discussed above. This may be described as a modified DCF hardcore
approach where the 'guaranteed' income or core income is discounted
at a risk free rate as there should be no uncertainty about its
payment (assuming a good covenant) and the most likely potential
uplift discounted at a risk adjusted rate reflecting the volatile
nature of this top slice income.

This approach seems well suited to highly geared short leaseholds
in particular where the income stream is limited and the method is
easy to use. But for reversionary investments (such as the example
used here) where the income stream is not terminable, the method
becomes unwieldly if the uplift income is separately calculated at
each review date in perpetuity. A manageable form of the sliced
income approach is shown in Table 6 where the income receivable until
the disposal date is split into two layers and the 'guaranteed'
income is discounted at a risk free rate and the uplift at each
review is discounted at a much higher risk adjusted rate.   The
disposal in year 12 is treated in the same way as in the RADR
approach.

A further problem with this method is in the choice of risk
adjusted discount rate; but where the uplift in income is not
expected to be very great this is not a significant problem. Where
uplift is significant some form of sensitivity analysis on the
valuation of the uplift would seem appropriate!

Table 6 Sliced income approach – assumptions and results

ERV		£24,000 pa
		(£15,000 pa 'guaranteed')
Rental growth 0-2 years		10% pa
2-12 years		8% a
Yield on disposal		4.5%
Capital Value		
Guaranteed Income	@ 10%	
Uplift Income	@ 16%	
Disposal	@ 12%	£502,000
	@ 13%	£468,000
	@ 14%	£437,000

Probability Analysis

Hillier Method

This method was first proposed by Hillier in 1963 and has been
modified and applied to property investment by a number of authors,
most recently in this country by Sykes:1983c and MacIntosh and

82

Sykes:1985. This model is relatively simple to use but gets progressively more complex as the number of variables increases and in these situations a Monte Carlo simulation approach would be preferable.

In essence the model calculates the expected net present value (NPV) in the normal way and then separately calculates the standard deviation of the NPV as illustrated in the simple example below (see Epley and Millar:1980 and Young:1977). Two alternative approaches to calculate the SD of the NPV are shown depending on whether the cash flows are assumed to be positively correlated or completely independent. In practice it may be that the answer will be somewhere between the two extremes and Young:1977 suggests an approach which deals with appraisals where part of the cashflow is independent and part perfectly correlated.

Table 7 Estimated cash flow example

Year	Type of Cash Flow	Expected Cash Flow (£)	Standard Deviation
0	Purchase	(125,000)	0
1	Income	10,000	0
2	Income	15,000	1,000
3	Income	20,000	1,500
3	Sale	250,000	20,000

It is assumed here that the cash flows are normally distributed and that the estimated extreme values of these cash flows represent $\pm$ 3SD's from the best or expected estimate.

For year 2 therefore it was assumed that the virtually certain cash flow was £12,000 compared to the best estimate of £15,000. Three SD's are therefore equivalent to £3,000 and 1 SD equals £1,000. For years 0 and 1 it is assumed that the cash flows are known figures, i.e. the purchase price is the price for which the property can be acquired and the initial rent has recently been agreed and so the tenant is contractually bound to pay it. For this simple example annual rent reviews have been assumed with a disposal at the end of year 3.

The NPV of the expected cash flow assuming a discount rate of say 12% is:-

$$NPV = -125,000 + \frac{10,000}{1.12} + \frac{15,000}{1.12^2} + \frac{20,000}{1.12^3} + \frac{250,000}{1.12^3} = £88,000$$

Where the cash flows are considered independent from year to year or period to period the SD of the NPV is given by taking the square root of the sum of the squared and discounted standard deviations of each cash flow.

$$\sigma^2 \text{ NPV} = \sum_{j=0}^{n} \frac{\sigma_j^2}{(1+j)^{2j}}$$

$$\text{SD}^2 \text{ of NPV} = \frac{(1,000)^2}{(1.12^2)^2} + \frac{(1,500)^2}{(1.12^3)^2} + \frac{(20,000)^2}{(1.12^3)^2}$$

$$= \frac{1,000,000}{1.574} + \frac{2,250,000}{1.974} + \frac{400,000,000}{1.974}$$

therefore, SD = £14,300

This analysis shows that the best estimate of NPV is £88,000 but that there is a 68% chance that the NPV will lie between £73,700 and £102,300 and a 95% chance it will be between £59,400 and £116,600.

Where the cash flows are considered positively correlated the SD of the NPV will be greater than where the cash flows are independent. Generally the greater the correlation the greater the dispersion of the distribution of the NPV and the greater the risk. The SD of the NPV is given by the sum of the present value of the SD of each cash flow.

$$\text{SD of NPV} = \frac{1,000}{1.12^2} + \frac{1,500}{1.12^3} + \frac{20,000}{1.12^3} = £16,100$$

As the example used becomes more complex and more relevant to a typical cash flow property investment appraisal such as the investment used in previous sections, the calculation becomes more difficult. The following problems must be addressed for a reversionary investment:

(i) The calculation of the standard deviation of each expected cash flow. The cash flows will be determined by a combination of the estimates of ERV and the estimates of rental growth and will be fixed for the periods between reviews. The calculation of the standard deviation is made more complicated as two variables – ERV and rental growth – are involved.

(ii) The calculation of the standard deviation of capital value when the investment is sold. The calculation of capital value involves three variables – ERV, rental growth and the market capitalisation rate. The calculation of the standard deviation is therefore a complex undertaking.

Where there are three variables, as there are with the reversionary investment example used in this paper, the Hillier method becomes unwielding and a Monte Carlo simulation, discussed later, would be more appropriate unless simplifying and potentially inacurate assumptions are made about the spread of likely capital values when the investment is sold. The extremes of this range of capital values could be taken as being ± 3 SD's from the most likely estimate, but this approach is very crude and is not to be recommended.

Where the investment is not reversionary the Hillier method is reasonably straightforward and this is illustrated in the following example where the reversionary investment used previously has been modified. It is assumed that the first review (after two years) has now occured and the rent payable is the actual rent that was achieved in 1987 namely £30,000 pa i.e. the investment is now rack rented. The remaining assumptions relating to estimated rental growth and capitalisation rate on disposal remain similar to those assumed previously except that the investment will now be held for 15 years. The assumptions are shown below:

	Period[1]	Best Estimate (a) %	Virtually Certain Range[2] (b) %pa	SD (c) %
Rental Growth	0-5 yrs	9	6 - 12	1
	0-10yrs	8.5	4.5 - 12.5	1.333
	0-15yrs	8	4 - 13	1.5
Capitalisation Rate (ARY)	15 yrs	4.5	4 - 5	0.1667

Notes

1   Rental growth for each period has been estimated from the commencement date (i.e. year 0 or today) rather than over each successive five year period using the previously forecasted rent as a base. This approach avoids the potential problem of correlation between successive cash flow estimates.

2   It is assumed that the variables are normally distributed so that there is a 99.7% probability of the values lying within the range of ± 3 SD's of the best estimate or mean, i.e. the virtually certain range is represented by ± 3SD's of the best estimate, so the SD is equivalent to one sixth of the value of this range.

The calculations that are now necessary are as follows:

1.   Calculate the present value of the forecasted cash flows.

2.   Calculate the variance (and hence the standard deviation) of the cash flows as illustated in the simple example above.

In the calculation of the variance the following points are important:

(i)   The income will be reviewed every five years so that the annual cash flow at the start of the cycle will remain the same throughout the five years, but each five yearly group of cash flows will be independent of the previous group of cash flows as rents are reviewed to the market rent at that time rather than the previous rent plus a percentage increase.

(ii) Income for the first five years will be certain. A range of forecasts is therefore not applicable and the variance and standard deviation will be zero.

(iii) The income at the start of the five yearly cycle will be given by the familiar expression:

$$R = R_o (1+g)^n$$

Where R = future rental value in n years time.
n = years until the start of the five yearly cycle
g = growth rate over n years

Where the standard deviation of the rental growth rate is σg the the standard deviation of the rental value for any five year cycle (σR) can be approximated by the following: (See Sykes:1983c or MacIntosh and Sykes:1985).

$$\sigma R = \frac{dR}{dg} \sigma g$$

which is equivalent to:

$$\sigma R = \frac{nR}{(1+g)} \sigma g$$

Inserting the actual figures shown earlier the standard deviation of the rental value in years 6-10 is:

$$\frac{5 \times 30,000 \times (1.09)^5}{1.09} \times 0.01 = £2118$$

Similarly the SD of the rental value in years 11-15 is:

$$\frac{10 \times 30,000 \times (1.085)^{10}}{1.085} \times 0.013333 = £8,333$$

Similarly the SD of the rental value at the date of disposal is:

$$\frac{15 \times 30,000 (1.08)^{15}}{1.08} \times 0.015 = £19,825$$

An alternative and possibly easier to follow way of calculating the SD of rental value is as follows:

Rental Value in Year 5:

Most pessimistic estimate = £30,000 x $(1.06)^5$ = 30,000 x 1.338
= £40,140

Most optimistic estimate = £30,000 x $(1.12)^5$ = 30,000 x 1.762
= £52,860

86

Therefore, maximum possible range = £12,720
Therefore, 1SD = £2,120

Rental values in years 10 and 15 could then be calculated in the same way.

The capital value on disposal will be dependent on two variables – the projected rental value and the forecasted capitalisation rate and this complicates the calculation of the variance. As both variables are independent, Sykes:1983 and MacIntosh and Sykes:1985 suggest the following approximation of the variance of the terminal capital value:

$$\sigma_v^2 = \left\{ \frac{dv}{dR}.\sigma_R \right\}^2 + \left\{ \frac{dv}{dy}.\sigma y \right\}^2$$

Where v = capital value
       R = forecasted rental income
       Y = forecasted capitalisation rate

As $v = \dfrac{R}{Y}$

Therefore $\dfrac{dv}{dR} = \dfrac{1}{y}$ and $\dfrac{dv}{dy} = -\dfrac{R}{y^2}$

So the equation can be re-written as:

$$\sigma_v^2 = \left\{ \frac{1}{y}.\sigma_R \right\}^2 + \left\{ \frac{R}{y^2}.\sigma y \right\}^2$$

Inserting actual figures for this example the equation becomes:

$$\sigma_v^2 = \left\{ \frac{1}{0.045} \times 19{,}825^2 \right\} + \left\{ \frac{95{,}160}{0.045^2} \times 0.001667^2 \right\}$$

Therefore v = £447,500

A summary of these calculations is shown in the following table together with the expected discounted value of the cash flows and their standard deviations which have been calculated using the approach described earlier.

Note:
_____

$\dfrac{dR}{dg}$, $\dfrac{dV}{dR}$ and $\dfrac{dV}{dy}$ are the partial derivatives

Yea	Cash Flow (Best Estimate) (£)	Estimated SD (£)
1	30,000	0
2	30,000	0
3	30,000	0
4	30,000	0
5	30,000	0
6	46,000	2,118
7	46,000	2,118
8	46,000	2,118
9	46,000	2,118
10	46,000	2,118
11	67,800	8,333
12	67,800	8,333
13	67,800	8,333
14	67,800	8,333
15	67,800	8,333
15	2,115,000	447,500

Discount rate   = 12%
Expected present value = £667,000
SD of present value   = £ 82,000

This rather more sophisticated risk analysis involving forecasts
and probabilities gives the investor a single point estimate plus the
additional information that there is a 68% probability that the value
will lie in the range of £585,000 to £750,000. Alternatively,
similar calculations can be undertaken to show the risk of the IRR
being achieved.

Because the resultant probability distribution of NPV's is a
normal distribution (as the returns are normally distributed),
additional information can be derived from the above calculations.
If the property was purchased for say £615,000 (see valuation on
p.74) this would show a positive NPV of £52,000 at a 12% discount
rate, with a standard deviation of £82,000 (as the purchase price is
a known figure its standard deviation will be zero and so the
standard deviation of the cash flow will be as calculated above).
The probability that the NPV will be less than £0 (i.e a 12% equated
yield will not be achieved) can be found by first calculating how far
£0 is from the mean or most likely NPV of £52,000 in terms of
standard deviations. If one standard deviation is £82,000 then £0
must be 0.634 standard deviations from the mean.

By using tables of normal curve areas (see Table 8 below) it is
fairly easy to calculate that the probability that the NPV is less
than £0 equals 26% (see Sykes:1983c or Young:1977, for a full
discussion of the characteristics of the normal distribution see
Byrne and Cadman 1984). This means that this investment has a 26%
chance of not meeting the equated yield target of 12%. This approach
is particularly useful when alternative investments are being
compared and provides additional information to assist in the
decision making process.

Areas under the normal curve

Z	0	0.01	0.02	0.03	0.04	0.05	0.06	0.07	0.08	0.09
0	0	0.004	0.008	0.012	0.016	0.0199	0.0239	0.0279	0.0319	0.0359
0.1	0.0398	0.0438	0.0478	0.0517	0.0557	0.0596	0.0636	0.0675	0.0714	0.0753
0.2	0.0793	0.0832	0.0871	0.091	0.0948	0.0987	0.1026	0.1064	0.1103	0.1141
0.3	0.1179	0.1217	0.1255	0.1293	0.1331	0.1368	0.1406	0.1443	0.148	0.1517
0.4	0.1554	0.1591	0.1628	0.1664	0.17	0.1736	0.1772	0.1808	0.1844	0.1879
0.5	0.1915	0.195	0.1985	0.2019	0.2054	0.2088	0.2123	0.2157	0.219	0.2224
0.6	0.2257	0.2291	0.2324	0.2357	0.2389	0.2422	0.2454	0.2486	0.2517	0.2549
0.7	0.258	0.2611	0.2642	0.2673	0.2704	0.2734	0.2764	0.2794	0.2823	0.2852
0.8	0.2881	0.291	0.2939	0.2967	0.2995	0.3023	0.3051	0.3078	0.3106	0.3133
0.9	0.3159	0.3186	0.3212	0.3238	0.3264	0.3289	0.3315	0.334	0.3365	0.3389
1	0.3413	0.3438	0.3461	0.3485	0.3508	0.3531	0.3554	0.3577	0.3599	0.3621
1.1	0.3643	0.3665	0.3686	0.3708	0.3729	0.3749	0.377	0.379	0.381	0.383
1.2	0.3489	0.3869	0.3888	0.3907	0.3925	0.3944	0.3962	0.398	0.3997	0.4015
1.3	0.4032	0.4049	0.4066	0.4082	0.4099	0.4115	0.4131	0.4147	0.4162	0.4177
1.4	0.4192	0.4207	0.4222	0.4236	0.4251	0.4265	0.4279	0.4292	0.4306	0.4319
1.5	0.4332	0.4345	0.4357	0.437	0.4382	0.4394	0.4406	0.4418	0.4429	0.4441
1.6	0.4452	0.4463	0.4474	0.4484	0.4495	0.4505	0.4515	0.4525	0.4535	0.4545
1.7	0.4554	0.4564	0.4573	0.4582	0.4591	0.4599	0.4608	0.4616	0.4625	0.4633
1.8	0.4641	0.4649	0.4656	0.4664	0.4671	0.4678	0.4686	0.4693	0.4699	0.4706
1.9	0.4713	0.4719	0.4726	0.4732	0.4738	0.4744	0.475	0.4756	0.4761	0.4767
2	0.4772	0.4778	0.4783	0.4788	0.4793	0.4798	0.4803	0.4808	0.4812	0.4817
2.1	0.4821	0.4826	0.483	0.4834	0.4838	0.4842	0.4846	0.485	0.4854	0.4857
2.2	0.4861	0.4864	0.4868	0.4871	0.4875	0.4878	0.4881	0.4884	0.4887	0.489
2.3	0.4893	0.4896	0.4898	0.4901	0.4904	0.4906	0.4909	0.4911	0.4913	0.4916
2.4	0.4918	0.492	0.4922	0.4925	0.4927	0.4929	0.4931	0.4932	0.4934	0.4936
2.5	0.4938	0.494	0.4941	0.4943	0.4945	0.4946	0.4948	0.4949	0.4951	0.4952
2.6	0.4953	0.4955	0.4956	0.4957	0.4959	0.496	0.4961	0.4962	0.4963	0.4964
2.7	0.4965	0.4966	0.4967	0.4968	0.4969	0.497	0.4971	0.4972	0.4973	0.4974
2.8	0.4974	0.4975	0.4976	0.4977	0.4977	0.4978	0.4979	0.4979	0.498	0.4981
2.9	0.4981	0.4982	0.4982	0.4983	0.4984	0.4984	0.4985	0.4985	0.4986	0.4986
3	0.4987	0.4987	0.4987	0.4988	0.4988	0.4989	0.4989	0.4989	0.499	0.499

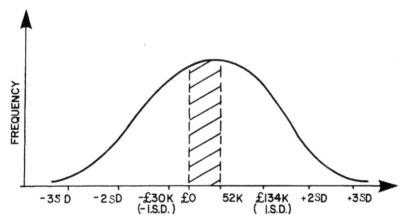

NPV of £0 represents $\dfrac{52,000}{82,000}$ = 0.634

From the table 0.634 S.D = 0.237 (i.e. the hatched area = 23.7% of the area under the curve)

Therefore, probability that NPV <£0 = 0.5 - 0.237 = 26.3%

The main problem with the Hillier approach is that for more complex appraisals with more than two variables (such as a reversionary investment) the method becomes unwieldy. Similarly some of the assumptions about the nature of the cash flows, e.g. that forecasts are normally distributed and all cash flows are either totally independent or completely correlated, are either restrictive or oversimplified. For a fairly straightforward non-reversionary investment the Hillier approach could well be a useful approach to risk analysis and is clearly an advance over the previous methods discussed but for more complete (and more realistic?) investment appraisals a Monte Carlo simulation approach would appear to be more appropriate.

Monte Carlo simulation

This approach has been used for many years in non-property investment appraisal and there are numerous text books and articles about its use. Recently, it has been put forward as a particularly useful approach in quantifying the risks involved in the appraisal of property developments. (See, for example the following UK texts: Bryne and Cadman:1984 and Darlow:1982 and 1988.)

Clearly, the large number of variables involved in a typical development appraisal makes a simulation approach very appropriate but it can also be usefully used for the appraisal of property investments where the number of variables make other methods unwieldly or over simplistic.

The method itself is very straightforward and easy to explain. The difficulty in its use relates firstly to estimating value ranges, and probability factors, for each variable (but this is true of any reasonably sophisticated method of risk analysis) and secondly, the large number of appraisal runs or simulations that have to be undertaken to enable a probability distribution to be formulated for NPV or IRR. A computer therefore is obviously essential.

There are two basic stages to Monte Carlo simulation:

1. The first involves estimating a range of values for each variable together with the likelihood or probability of each value occurring. In its most simple form this could mean assuming that the values for each variable were normally distributed so that only the most likely value and the virtually certain range need to be estimated (as described for other approaches in earlier sections of this chapter). But the advantage of simulation is that this constraint need not apply and skewed distributions can be assumed if they are thought to be more appropriate.

2. The second stage involves undertaking an appraisal with randomly selected values for each variable. This process is then repeated over and over again, each run producing an NPV or IRR. Because some values for each variable have a greater probability of being achieved than others, random selection will mean that they appear more frequently in the numerous appraisals undertaken. Sometimes different combinations of each variable will produce the same NPV/IRR, so that a pattern builds up of NPV's/IRR's and their chance of occurrence. This pattern can be graphically portrayed in different ways and will show the investor how likely certain values are to be achieved.

Typically, the following information would be shown as a result of numerous runs (absolute minimum of 100). This is similar information to that produced by the Hillier method discussed in the previous section.

(i)   Best esimtate of NPV/IRR (not necessarily the same as calculating NPV/IRR by using the best estimate of each variable).

(ii) The standard deviation of NPV/IRR.

(iii) Ranges of NPV/IRR and their probability of being achieved.

(iv) The cumulate probability of achieving or not achieving a certain NPV/IRR.

Table 8 gives an illustration of the estimates that could be made using the reversionary investment described at the start of Section 4.

Table 8

Variable	Range	Probability %	Probability Number (out of 100)
ERV	22,000	10	1-10
	23,000	30	11-40
	24,000	35	41-75
	25,000	20	76-95
	26,000	5	96-100
Rental Growth	8% pa	10	1-10
Years 0-2	9% pa	35	11-45
	10% pa	40	46-85
	11% pa	10	86-95
	12% pa	5	96-100
Years 0-7	6% pa	15	1-15
	7.5% pa	25	16-40
	9% pa	35	41-75
	10% pa	20	76-95
	11% pa	5	96-100
Years 0-12	4% pa	15	1-15
	6% pa	30	16-45
	8% pa	35	46-80
	9.5% pa	15	81-95
	11%	5	96-100
Capitalisation	4% pa	10	1-15
Rate	4.25% pa	25	16-30
(on disposal)	4.5% pa	35	31-65
	4.75% pa	30	66-93
	5% pa	5	96-100

Further refinements to the assessment of values for each variable are possible although the programme will be more complex and take longer to run. One refinement would be to take more values within the parameters of the range adopted, i.e. say ten values rather than the five that have been assumed above. Alternatively a limited range of values for each estimate rather than a single point estimate could be assumed, with the computer randomly selecting a value within that range for each run. For example, the ERV the table might be refined as follows:

Range	Probability %
21,500–22,500	10
22,501–23,500	30
23,501–24,500	35
24,501–25,500	20
25,501–26,500	5

Further modifications could be made by adding in other variables such as refurbishment, voids when leases end, etc., as deemed appropriate.

The advantage of this type of risk analysis as with the Hillier method is not just the greater amount of information that an investor has to enable a decision to be made but the ability to compare different possible investments on a range of criteria other than just the best estimate of NPV/IRR. The following examples illustrate this.

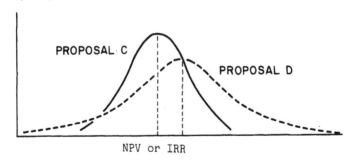

NPV or IRR

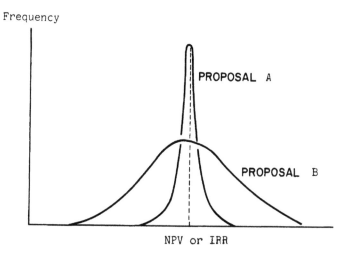

Frequency

PROPOSAL A

PROPOSAL B

NPV or IRR

Project A and Project B have the same best estimate of NPV and on a one point DCF appraisal would rank equally but an examination of their risk profile suggests Project B has a larger standard deviation and is therefore more risky. For Project B there is a greater chance of achieving a low NPV/IRR and a greater chance of achieving a high NPV/IRR than in Project A but a lower probability of achieving the best estimate of NPV.

A comparison between Projects C and D is more difficult as Project C would be described as being less risky but also not achieving such a high best estimate of NPV/IRR as Project D.

The decision as to which investment should be acquired will depend on the investor's attitude to risk and the investor's overall portfolio (see Chapters 7 and 8). However, in general terms, one would expect Project A to be preferable to Project B.

Other useful methods of comparison between investments include the risk profile approach, the probablity of loss approach and the liquidity approach (see Phyrr:1973).

In the risk profile approach, rather than setting the probability of receiving a specific rate of return on the vertical axis, cumulative probabilities are shown in descending order and the IRR plotted on the horizontal axis. Investments can then be compared for any level of IRR to see which project has the highest probability of exceeding that IRR. It may be that investments will rank differently depending on the level of IRR required.

The probability of loss approach involves computation of the expected IRR and the probability of not achieving a required IRR as discussed earlier on p.83. These results can be portrayed graphically and different investments compared. Investments can be eliminated that are not likely to achieve a specified IRR and an investment can be chosen which produces the highest IRR for any given level of risk or the lowest risk for any desired IRR.

The liquidity approach involves an examination of cash flows, identifying those that are negative, so that for example investments are eliminated which generate a certain number of successive negative cash flows even if the overall IRR is acceptable. This approach is more applicable to development appraisal than investment appraisal but could be useful where voids and capital expenditure are possible in certain years.

In certain respects Monte Carlo simulation would appear to be an advance over the Hillier method although the use of a computer is absolutely essential. The major problem, as with the Hillier approach, concerns the possible dependency between variables. For example in the reversionary investment appraisal used in this section it is probably true to assume that ERV and rental growth are not correlated so that the rent at subsequent review dates could be any combination of ERV and rental growth rather than a high estimate of ERV being linked with a high estimate of rental growth and vice versa. But is it equally likely that rental growth up to the disposal date and the capitalisation rate at the disposal date will not be correlated? It is perhaps more likely that high rates of rental growth preceding the disposal date may influence market sentiment and hence mean that a low capitalisation rate, or investment yield, is more likely than a high one and vice versa. Over time there should in theory be some correlation but accurately assessing it is another matter.

The model can be adjusted so that different ranges of values for certain variables are linked together (see Byrne and Cadman:1984) but the problem is not so much in adjusting the computer program to accommodate this refinement but in accurately assessing the linkages. It is hard enough assessing an accurate range of possible values for each variable, harder still putting meaningful probability factors to each value within that range, but stretching the bounds of acceptability to try and link certain values of one variable with certain values of another variable. Theoretically it is possible, but practically, it seems some way off. Phyrr:1973 is of the opinion that the use of such complex Monte Carlo simulation models will 'depend to a substantial extent on industry acceptance of probabilistic models for planning and decision making, and on management's ability and willingness to forecast and estimate probability distributions for uncertain variables. It will perhaps, take a considerable orientation period and a thorough understanding of probability theory before confidence in such a model will evolve in real estate'.

5 Conclusion

A variety of different approaches to risk analysis have been presented in this chapter which have been devised or developed by a variety of academics and practitioners over the past 25 years. Few, if any, are currently widely or even occasionally used in practice for investment appraisal in this country. The methods encompass (i) simple sensitivity analysis where a range of values rather than a single estimate are provided (but with no explicit probability attached to this range), (ii) simple deterministic estimates crudely taking account of probability and risk; and finally (iii) models which provide a single best estimate but also some measure of the probability of a range of values.

The cruder the method the easier it is to use in practice but the less advance it provides over standard DCF methods. The more sophisticated the method the more difficult it is to use in practice mainly due to the difficulty of estimating the range of values for different variables. There is a great advance in the amount of information that is provided by these models but the confidence that investors will attach to the results will depend on the confidence they have in their estimates for each vaiable. To quote Phyrr:1973 again:

> 'These adjustments to allow for uncertainty may be challenged as nothing more than guesses. Perhaps they are, but even so, they are guesses that must be made and will be made, either explicitly or implicitly. Failure to apply the probability adjustment does not avoid the problem, it merely transfers the guess element in a disguised form to some other stage of the decision making process.'

References

Baum A:1987, Risk Explicit Appraisal: A Sliced Income Approach. Journal of Valuation.

Byrne P and D Cadman:1984, Risk and Uncertainty in Property Development. Spon.

Darlow C (Ed):1982:1988, Valuation and Development Appraisal, Estates Gazette.

Epley D and J Millar:1980, Basic Real Estate Finance and Investments. John Wiley and Sons.

Hager D and D Lord:1985, The Property Market, Property Valuations and Property Performance Measurement. Institute of Actuaries.

Hillier F:1963, The Deviation of Probalistic Information for the Evaluation of Risky Investments. Management Science Vol.9, April.

Jones Lang Wootton:1987, Obsolescence. The Financial Impact on Property Performance, JLW.

MacIntosh A and S Sykes:1985, A Guide to Institutional Property Investment. MacMillan.

Miles J:1987, Depreciation and Valuation Accuracy, Journal of Valuation.

Phyrr S:1973, Computer Simulation Model to Measure the Risk in Real Estate Investment. The Real Estate Appraiser May-June.

Salway F:1986, Depreciation of Commercial Property. CALUS.

Sykes S:1983a, Property Valuation, Investment and Risk; in Land Management - New Directions Edited by Chiddick and Millington. Spon

Sykes S:1983b, Uncertainties in Property Valuation and Performance Measurement. Investment Analyst, January.

Sykes S:1983c, The Assessment of Property Investment Risk. Journal of Valuation.

Young M:1977, Evaluating the Risk of Investment Real Estate. The Real Estate Appraiser.

CHAPTER 6

FORECASTING OF RENTS

1   Forecasting Property Values

Forecasting is not new to the property scene. A valuation of a
building involves an element of forecastng. A valuation based on an
all risks yield is making an assumption about future rental growth
for the building being valued. If this was not the case a cost of
money yield with perhaps a premium to take account of the illiquidity
of the investment would be used to value the building. Likewise a
property consultant advising on the purchase of a new property and
looking for a bargain for his client must be making some assumptions
about the future. These assumptions, as in the case of the market
when setting capitalisation yields that are less than money rates,
may not be explicit - 'gut feeling' is often referred to - but
forecasts they are nevertheless. And there are many other examples
of forecasting. This is inevitable where decisions must be taken
today that will affect future performance.

What is relatively new is formal forecasting based on
mathematical models. The sheer size of the property investment
market and the amount of money involved requires a formal approach to
decision-making. There is a growing opinion that advocates this
approach to valuations and specifically the use of discounted
cashflow methods. These involve rental growth projections although
this is strictly the next stage on from forecasting, namely at some
point a forecast on rental movements is made and these are then input
into the discounted cashflow valuation model.

It is worth considering for a moment why we need to forecast at
all. Put very sinply if an investor is completely ignorant of future
movements in the investment markets then he is unlikely to be able to
come to any meaningful decision. These decisions can be at the macro
level - should one be investing in property at all? - or at the
micro level - what rental growth can I expect from property X? It is
not necessary to predict exact levels when forecasting. It is more
important to detect turning points. When is the market going to turn
up? Which sector of the market is going to perform best? Forecasting gives insight into these questions, the answers to which are
essential for successful investment.

There are five basic indicators or variables in measuring
property performance:

    Yield
    Rent
    Rental value
    Capital value
    Rate of return

We will be concentrating on the first and third of these
variables as once they have been forecast the others follow
naturally. Before looking at the actual forecasting process there is
one further distinction that I will be making. That is the

difference between forecasts of market variables and forecasts of portfolio variables.

2 Market and Portfolio Variables

A property portfolio will have been built up over a number of years and will contain buildings of varying ages, some with long lease review structures, some having just been reviewed and others will be at different times between reviews. The performance of the portfolio will also be affected by factors such as active management and depreciation. Different portfolios although having similarities will vary significantly in structure and how they have been managed. Forecasting portfolio performance will depend very much on the actual structure of the portfolio being analysed and will not necessarily translate into all other portfolios.

Market variables are a reflection of today's situation and react very quickly to a changing economic climate. They measure activity in the market. In particular rental values, or the movement of rental values, are a reflection of tenant demand and the level of supply. Likewise investment yields react to investment demand, that is the amount of money being invested in property. In many ways market movements are a forerunner of portfolio movements. A simple example of this is that rent reviews are rarely annually and more often every five years. Portfolio rents consequently will not immediately reflect current market rental values. Portfolio capital values will also not immediately react because of the rent review frequency.

Both market and portfolio variables are important measures of property as an investment. Lack of space prevents me from looking at the forecasting of both sets of variables. I will concentrate, therefore, on market variables in the rest of this chapter.

3 Supply and Demand

The movements of rental values and yields are classic examples of the laws of supply and demand. In the case of rental values demand comes from tenants looking for floor space to extend their business. We have seen a prime example of this in the office market over the last 15 months. The expansion of the stock markets with business volumes reaching record levels in the lead up to 'Big Bang' and the following months resulted in high tenant demand for large dealing floors in the City of London. Lack of adequate supply resulted in prime rentals rising from about £30 per sq ft to over £60 per sq ft in a very short period of time. Rentals in the City rose on average by over 45% during 1987. The crash in the markets in October forced the banks and brokerage houses to stand back and take stock of the situation. This resulted in demand moderating and a slowing in the pace of rental growth.

This is an example of a specific sub-market in the office sector responding to demand restricted to that sector. But rentals can also be affected nationally. The retail market has been a prime example of this during 1987. Figure 1 shows the movement in retail sales volume and retail rents adjusted for inflation. Here we see rental

Figure 1   Trends in Shop Rent 1965–1988:   Inflation Adjusted

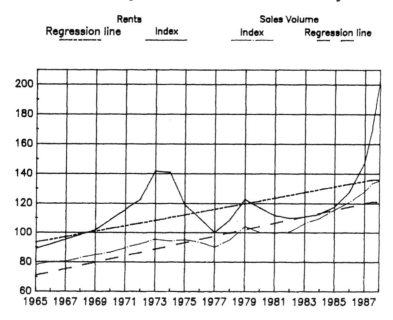

Figure 2   Trends in Industrial Rents 1974 to 1987

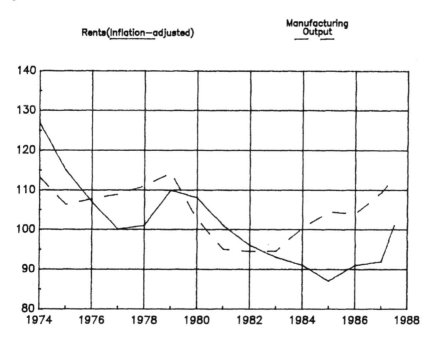

growth following very closely that of retail sales volume. It did, however, encourage retailers to expand and new players to enter the retailing market. Retail sales is a proxy form for this expansion in floor space demand which pushed rents up.

A further illustration of this has occurred in the industrial market during 1987 and particularly the second half. Industrial rents showed very little growth for much of the 1980's and it is not difficult to see why. Inflation adjusted industrial rentals are plotted against manufacturing output in Figure 2. Manufacturing output peaked in 1979 falling dramatically until 1981 at which point it plateaued out. At the same time industrial rents fell from a peak index value of 110 to 101 and continued falling until 1985. Output actually started to rise from 1983 and has continued to do so into 1987 where it stands just below its 1979 level. Rents responded in 1985 but it was not until 1987 that the increase was significant. Figure 2 is interesting for one other factor. The peaks of the two series appear to be co-incident whereas it is about two years after an upturn in output that rents turn up. The series are too short to draw definite conclusions but they do suggest a pattern.

Again it is not manufacturing output that has moved rents but rather manufacturers' need to increase floor space in order to produce the extra output needed to support the consumer boom.

The above illustrates important steps in the forecasting process. Demand is a key variable in forecasting rentals. Demand however is very difficult to quantify and predict and it is often necessary to look for proxies. These proxies of course have to be reasonable. Therefore it is necessary to identify variables which bear some relationship to tenants' need to expand. The above examples also illustrate one further aspect of forecasting. The last two examples demonstrated repeatable demand. That is retailers will respond regularly to increases in sales. The City of London example demonstrates a one-off situation which will not be repeated in exactly the same form. Needless to say the one-off situation is much harder to forecast.

Property yields are also a function of supply and demand but in this case demand comes not from tenants seeking to occupy a building but from investors. This is much more difficult to quantify. Yields fall (i.e. capital values rise) as the amount of investment increases and rise as this decreases. It is not the only influencing factor but is probably the most important as illustrated in Figure 3. This shows investment in direct property by institutional investors between 1973 and 1987.

Investment tends to be volatile so Figure 3 uses a four-quarters moving average. Institutional investment rose strongly over the period 1979 to 1981, remained at those levels but then fell away sharply from 1983. Yields had been falling since 1974 during a period of high inflation with rental growth since 1977. The massive surge of investment in 1979 kept yields at historically low levels despite the inflation rate falling and the economy in a recession. When the investment funds fell away in 1983 yields started to move upwards. Clearly yield levels are a function of investment and this will be looked at in more detail later.

Up to this point we have concentrated on the demand side of the

Figure 3 Institutional Investment in Property 1973 to 1987

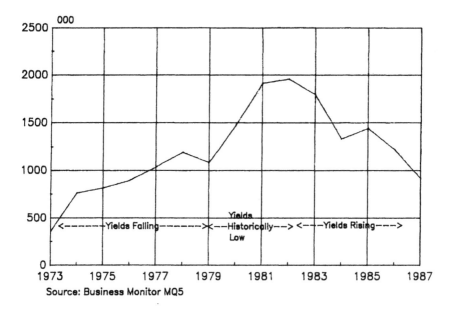

4—quarters moving average

Source: Business Monitor MQ5

Figure 4 Relationship Between Adjusted Shop Rents and Rental Profits (lagged six months) 1971–1981

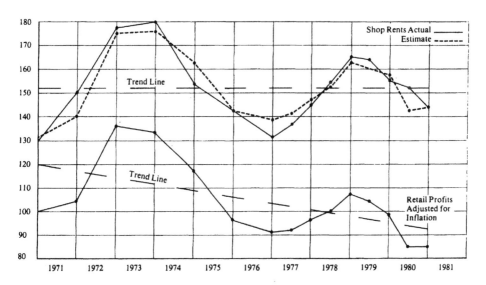

equation.   Supply is also important as even during periods of high
demand :f the supply is abundant then this should moderate any rental
rise or yield fall which would have been sharper if supply was in
short supply.   Except in specialised markets supply is much more
difficult to quantify.   The City of London office market is one such
example where good meausres are available.   In contrast, there is no
accurate measure of the national supply of retail space.   Measures of
empty industrial floor space are available but how much of this
represents the type of building tenants are looking for is open to
doubt.   However, looking into the supply side of the equation only a
marginal effect from national trends is found.   In more specialised
markets supply becomes much more important.

4.   Forecasting Rental Values

Firms such as Hillier Parker have been forecasting shop rental values
on a national basis since 1977.   As a measure of shop rents the ICHP
Rent Index covers the period 1965 to 1987.   Using this series a
simple model can be estblished between shop rents and retail sales
volume.   Various methods exist for quantifying relationships and we
used simple regression analysis.
     Retail sales volume is a measure of sales value having taken
account of inflation.   We therefore used inflation-adjusted shop
rents in the model.   At the time of our first forecast the series
only covered a period of 12 years and included only one cycle.   The
conclusions we drew at that time could only be tentative.   Despite
this sales volume turned out to forecast correctly the upturn that
occured after 1977.
     The same model proved adequate in 1978 but by 1979 with the
series producing two cycles it became obvious that retail sales
volume was no longer suitable.   The problem arose and we can see this
in Figure 2 because the cyclical movements of the rent series are
much more volatile than those of the sales volume series.   As the
chart clearly shows the turning points of the two series up to 1979
were still co-incident.   We concluded from this very simple analysis
that we could still use retail sales volume in our equation but that
some other variable or function was needed to compensate for the much
greater cyclical movements of rents.   We looked at a number of
possible variables but none reproduced the type of cyclical movements
we were seeking.   We resorted to using a dummy variable and after
some analysis came up with a cosine function which produced an
exceptionally good fit to the data.

Rents = 1.393 x (Retail Sales Volume) + 14.362 cosine (/3.25)

     The model worked reasonably well predicting correctly the coming
down-turn but being about six months out on timing.   One of the
problems turned out to be the forecast of sales volume used.   We have
tended to use the average from a number of sources and in the event
these turned out to be too high.   When we substituted the actual
sales volume figures the model proved more accurate.

This encouraged us to use the same type of model in our next forecast although it is never entirely satisfactory to use dummy variables. After the problems encountered with the previous forecast for retail sales volume we also produced a sensitivity table showing the effects on our forecast for different levels of retail sales volume. This produced our most pessimistic forecast of the four so far produced. We expected rents to grow just ahead of inflation during 1980 but we could not see them regaining the ground lost in the previous 12 months.

Even this forecast turned out to be too optimistic and we decided that the trigonometric dummy variable was not appropriate. Also despite sales volume holding up in what had become a deep recessionary period shop rents were falling quite steeply in real terms. We can see from Figure 1 that sales volume started to rise in 1982 but there was no response from rents. It had become obvious that some fundamental change had occurred in the market. Various other retail indicators were examined together with a return to look at the supply side of the equation. On this question we used our records of shopping schemes of over 50,000 sq ft which had opened since 1965. We had been monitoring those for some time and by 1981 had a reasonably long time series. As in previous attempts the supply side of the equation did not produce any significant results. Various other variables singly and together were considered. Among these were consumer expenditure and earnings. Of the variables considered store profits showed the best potential. There were, however, two main problems with the series. Store profits are highly seasonable with the bigger profits occurring round Christmas time and consequently highly volatile. The second problem was caused by the fact that the trend line for shop rents was rising slightly while that for profits (after adjusting) was declining.

The first of these problems was solved to some extent by using four quarters moving averages. The second was solved by removing the trends from the two series and looking at the residuals only. The procedure used in forecasting rental growth then took a number of steps.

Step 1: A new series for the retail profits was calculated by first deflating the profit growth figures by the retail price index. An index was then formed from this new series by taking the average of the previous four quarters, growth and chain-linking these together.

Step 2: Trend lines for both the rental and profits series were then calculated by regressing the series against time.

Step 3: The differences between actual and trend lines were then calculated at each time point.

Step 4: The residuals from the two series were then correlated using the method of least squares which produced a correlation co-efficient of 0.94. This is extremely high for this type of data.

Step 5: Using independent forecasts of retail profits, forecasts for shop rents above/below trend were arrived at. The trend line was then extrapolated and the forecast residuals added/subtracted to arrive at the final forecast.

The results of this process are shown in Figure 4. This produced a forecast of rents rising at a rate just below inflation. This proved reasonably correct. Again when we assessed the model after

103

the first year of the forecast period using actual profits the results were closer to the actual out turn.

Despite the good results from this model our aim was to find one or more variables which would explain or compensate for the different trends in the rent and profit series. This brought us back to square one turning again to retail sales volume. Using regression analysis again we established a relationship between shop rents adjusted for inflation, retail sales volume and a smoothed series of profits lagged six months. This produced a correlation coefficient of 0.93 indicating a reasonable fit. Despite the fact that the model fitted the historical data reasonably well it became evident over the next two years it was becoming increasingly divergent from the actual rent data. This led us to examine the type of property covered by the ICHP Rent Index.

The Rent Index covered the highest rented areas in 70 centres. Not all of these centres are considered prime by the investment market, but the Index generally represents good town centre shop property of the type mainly preferred by institutional investors at that time. It was becoming apparent that despite good profits and sales this type of property was not responding in terms of rental growth. This led us to examine other types of retail property and in particular to look at rental growth in towns of varying size.

We split our sample into five groups: the very largest cities (e.g. Birmingham), the next level down (e.g. with less than 20 multiples) and central London. Indices for each group were then regresse against retail sales volume producing some significantly different resul In terms of relationships measured using the correlation co-efficient th deteriorated inversely with size of town:

	Correlation
Central London	−0.32
Top 15 Cities	0.07
Middle 13 Towns	0.31
Lower 12 Towns	0.41
Smallest 22 Towns	0.97

These results suggested at the time that shop rental growth trends vary significantly between towns of varying size. Examination of the data further showed that this separation occured around 1981 which turned out to be the nadir of the recession. Obviously to produce an overall forecast of institutional shop property rent was becoming increasingly difficult. Further analysis showed central London rentals correlated very closely to the volume of overseas visitors to London while small town growth ran almost parallel to sales volume. In between the picture was less clearcut. This obviously required further research which we are now undertaking. In the meantime we have continued to forecast an overall rental growth using the same model but replacing retail sales volume by the slower growing real disposable income.

The procedures described and the general progression have subsequently also been successfully applied to office and industrial rents.

5 Yields

Determining suitable variables that explain demand is the most
difficult part of forecasting. Yield movements are a function of
investment demand. This will depend on several factors:

- how property is performing relative to other investments
- investors' perception of future rental growth
- the need to build up a suitable portfolio weighting in an
  important investment sector
- the perception of balancing risk and return
- the urge to invest in a real asset which becomes wholly owned

There are no doubt other reasons but these serve to illustrate
just how complex it can be in deriving a quantitative measure of
demand. We can measure this demand quantitatively through
institutional and corporate buying of property. What motivates this
is extremely difficult to quantify.

Supply depends to a large extent on developers' perception of
current and future growth in lettings which must in the end depend on
the health of the economy. The recent buoyancy in the level of
retail sales, for example, has produced a huge upsurge in the
development of retail space.

The research which we have carried out at Hillier Parker has
attempted to encapsulate these varying factors into quantifiable
equations which can be used to explain the yield movements over the
last 14 years. Many of these factors are highly volatile – much more
so than the movement in property yields. The equations are therefore
of principal value in detecting changes in direction of yields rather
than what their absolute level will be.

They have substantial uses among which are:

- detecting changes in the direction of yields
- forecasting whether yields will be higher or lower in the future
- identifying periods when yields are too high or too low

As with rental forecasts an index or measure of yields is
required. For our models we have used our average yield index which
covers the period since 1972.

6 Yield Trends and Interest Rates

The movements of these average yields is shown in Figure 5. The
figure shows how yields rose sharply between 1974 and 1975 as
confidence wilted following the collapse of the secondary banking
market. There was also a sterling crisis at the time and interest
rates went to a then historically high level. Yields could only go
one way particularly in an almost non-existent market place and that
was up. As confidence returned yields gradually fell reaching an all
time low point in 1979 and remained on this low plateau until
mid-1981 despite again record levels of interest rates.

This was an important period because as the figure clearly shows
up to 1979 there existed a reasonable elasticity between interest

rates and property yields. 1979 represented a breakdown of this linkage. By late 1981, however, property yields were again moving up and have continued doing so. Perversely, interest rates have been on a downwards path over this period. By the end of 1986, however, property yields were almost back to their 1974/75 levels.

What we have experienced then over the last 15 years is an apparent disentangling of the tie between interest rates and property yields with 1979 being the year of change. Another explanation is of course that the linkage still exists but certain vibrant external factors which previously had a neutral effect became dominant post 1979.

This is reasonable because interest rates represent the 'safe' investment and consequently behave as a target that other investment media must at least match if they are to attract investment money. If interests rise then so must this target, all other things being equal. In the case of property if it is perceived that rents will not grow at the required rate to justify prevailing yield levels then these must rise to compensate for this. Any linkage is consequently not necessarily a direct one.

It is, therefore not unreasonable to expect some degree of elasticity between property yields and interest rates. The plot of Figure 5 suggests that this state of events certainly existed up to 1979. To check this we correlated property yields against gilt yields over the period 1972 to 1979 only. The result is shown by the broken line in Figure 5. The correlation proved quite significant. Using the resulting equation we then extrapolated property yields over the period 1979 to 1987. What this line shows is how property would have moved if the assumed pre-1979 relationship with gilt yields had continued beyond then.

The difference between actual and projected yield movements is shown quite clearly. Thus on the basis of interest rate levels:

- property yields were just about right between 1972 and 1979
- between 1979 and 1982 property yields were too low, i.e. property was overvalued
- since 1982 property yields have been too high, i.e. property has been undervalued

7   Investment in Property

There were two main reasons for the very low levels of yields between 1979 and 1982. In the two years prior to 1979 inflation was at very high levels fuelling growth in rents. During that time returns from property easily beat that from other investment sectors. This attracted the institutions into the market - particularly pension funds who were very much underweight as far as portfolio holdings were concerned. This produced a huge inflow of capital into the market between 1979 and 1982 as shown in Figure 3.

Consequently, despite inflation being brought down to much lower levels and rental growth slowing perceptively, property yields were stuck down at historically low levels. And this against a background of rising interest rates which moved to their highest levels both in nominal and real terms.

106

Figure 5   Comparison Between Gilts and All Property Yields

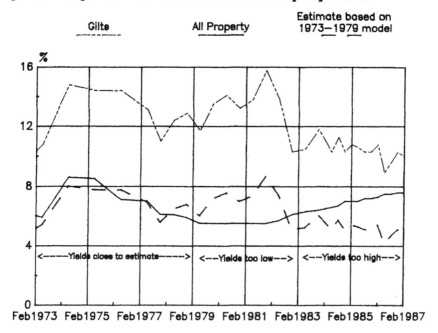

Figure 6   Comparison of Investment and Yield Residuals

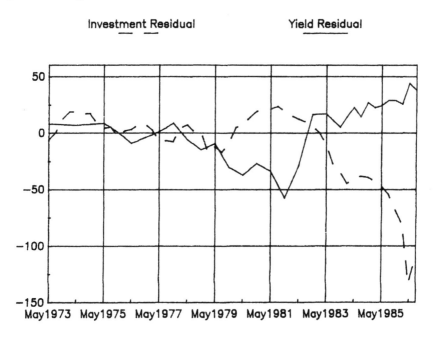

Institutional investment is, of course, not the only source of
money going into the direct property market. Property companies also
represent a very large section of the market. They raise money via a
number of vehicles, e.g. debentures, rights issues, but one of their
highest sources is through borrowings from the banking system. This
has been highly cyclical over the period of measurement with very
high levels of borrowing 1973/74 and since 1984. Between 1974 and
1979 there was a high level of repayment. The investment patterns
are distinctly different from those of the institutions. Calculating
the long term trend in the investment patterns of institutions and
property companies highlights the following:

	Property Companies	Institutions
1973 to 1979	Higher than trend up to 1974. Below trend to 1979	Close to long term trend
1979 to 1982	Below trend to 1981, Close to trend in 1982	Much higher than trend
1982 to 1986	Higher than trend	Lower than trend

We examined the differences between institutional investment and
its long term trend and the difference between actual yields and
those predicted by the model shown in Figure 5. The results are
given in Figure 6. This shows the two series as being a mirror image
of the other. This suggested to us that yield movements could be
described adequately by the movements of three variables:

-   gilt yield on high coupon, long dated stock
-   institutional investment in property
-   bank lending to property companies

The first of these (gilt yields) provide the target which
property returns must achieve to be attractive for investment.
Investment provided the demand. In fact we used the residuals of the
investment about their trend lines in the equation.

To test the adequacy of this we correlated property yields with
the three explanatory variables and the results are shown in Figure
7. This yielded a correlation coefficient of 0.9. In other words
over 80% of the changes in property yields can be explained by
changes in the three explanatory variables.

The model produced quite accurately defines the turning points in
the yield movements. The model can be seen to be more volatile than
the actual series and further smoothing of the input variables may be
necessary.

Figure 7   Models of Average Yields

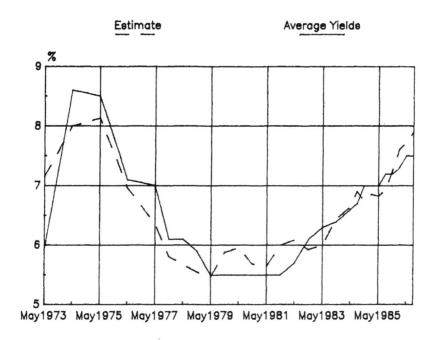

8. Conclusion

I have very briefly described a simple technique for forecasting
property rentals and yields. The most difficult part of any model
building concerns getting adequate explanatory variables. As we have
seen from our experience with forecasting rental movements these can
even change over time.

CHAPTER 7

PORTFOLIO THEORY AND PROPERTY INVESTMENT ANALYSIS

1   Introduction

Property is part of an integrated financial market and efficient
decisions must be made within this context. This view is important
because property has to compete with other sectors of the market when
it comes to allocating resources.
    To make significant advances in understanding property it is
essential to draw on the rich vein of investment research which has
been undertaken in other markets over the last thirty years. This
research has had a significant effect on the way investors behave and
if applied properly could have the same impact on the property world.
    There can be little doubt that the single most important factor
affecting investment has been the development of portfolio theory.
It is significant because it is at the root of our understanding of
the way assets should be priced, how resources should be allocated
and the way performance should be measured and analysed.
    The property profession in ths country has been slow to respond
to these ideas not only because of the high level of mathematics
involved but also because property has in the past been regarded as
outside mainstream investment and therefore required a special
approach. Fortunately, this position is changing and will no doubt
continue to do so as pressure is placed on funds to achieve superior
performance.
    This chapter presents a non-technical overview of the most
significant develpments in portfolio theory and demonstrates their
importance to the property sector.   It is considered in three parts.
    Part 1 covers the background to portfolio theory and its
development into capital asset pricing. It stresses the importance
of market risk and shows that investors are not compensated for all
the risk they take on.
    Part 2 develops the ideas in Part 1 and shows that rational
investors will wish to maximise net present value and that this
approach is wholly consistent with the concept of utility
maximisation. The correct valuations of assets are shown to be
present values and the quality of a valuation depends on the
available subset of information.   The implications of this approach
on market efficiency are also discussed.
    Part 3 addresses the problem of asset selection in relation to
the notion of maximising net present value. The limits of
diversification obtainable from a property portfolio are discussed
and demonstrated with some empirical evidence. The implications for
the measurement of portfolio performance are also discussed.

PART 1

Up until the early 1950's the notion of diversification as a means of reducing risk was treated subjectively. Faced with the problem of constructing a portfolio investors would address the risks involved merely by examining the qualitative aspects of the assets being considered. There was no recognition that the price of an asset should be determined by its contibution to a highly diversified portfolio.

These concepts were developed over the following decades but were founded in the seminal work of Harry Markowitz. His major contribution to finance was to show that every asset could be described in terms of its mean and standard deviation of return and that diversifications depended on the way returns moved relative to each other. He developed a complex algorithm which estimated the proportion of funds to be invested in each asset in order to maximise returns for given levels of risk or minimise risk for a given return.

Up until this time (the early 1950's) the proportion of funds to be invested in individual assets had no theoretical basis. Markowitz showed that not only was this problem capable of solution but it was also possible to show that a Markowitz portfolio would actually increase in risk as more assets were added. This was the opposite of conventional wisdom and clearly showed that there were significant gains to be made from using his approach.

Although his proposal was revolutionary it was not readily accepted, largely due to the fact that it involved a level of mathematics which was not commonplace at that time. The technique also required the use of computers which in those days were very much in their infancy. Although these were stumbling blocks the principal difficulty in its use, however, lay in the fact that the right type of data were not available. Expected returns, variances and covariances were not the sort of numbers which were readily available. Consequently the implications of his approach lay dormant for almost a decade.

The next major breakthrough in portfolio theory and its application to real problems came in the early 1960's with the development of the Sharpe model. This was a technique which simplified the calculation of expected returns, variances and covariances by relating the returns on any asset to some common index. Paradoxically the suggestion for this approach was contained in a footnote in a book published by Markowitz several years earlier.

The model developed by Sharpe enabled all the data for a full portfolio analysis to be estimated with relatively few inputs. Computer time could therefore be reduced and as a result efficient portfolios could be constructed with relative ease. At that time computing costs were a major consideration. This of course is no longer the case.

What was significant about the Sharpe approach was that it encouraged academics to think about the implications of portfolio theory in a wider context. Sharpe was amongst the first to consider the implications of all investors adopting the rationale of Markowitz. The results of this research gave rise to what has become known as Capital Market Theory.

111

Briefly this developed from recognising that the opportunity to lend or borrow at the risk free rate would enable portfolios to be constructed which lay outside the efficient frontier as defined by the opportunity set. This is shown in Figure 1. The straight line is known as the Capital Market Line and is the true efficient frontier. Only portfolios which are efficient in the Markowitz sense can lie on this line. All other portfolios and individual assets will lie below the line.

The importance of this concept is that portfolios which lie on the Capital Market Line do so only after the process diversification. Such portfolios will have a minimum level of risk which cannot be diversified away. This implies that the total risk of a portfolio is made up of two components. One reflects the level of risk which through the process of diversification can be eliminated and the other reflects the level of risk which cannot be diversified away. In Capital Market Theory these are referred to as the 'unsystematic' and 'systematic' risk components. The total risk on any portfolio can therefore be represented as the sum of these two components such that:

total risk = systematic risk + unsystematic risk

These terms often have different expressions applied to them. For example, systematic risk is often referred to as 'market risk' as it is largely determined by market effects whereas unsystematic risk is often referred to as 'non-market' or 'specific' risk. In other words the unsystematic component of risk is a function of specific factors relating to individual assets.

As the unsystematic or specific risk can be eliminated through diversification it will be evident that the element of risk which assumes greatest importance is systematic or market risk. It assumes importance because it is this element of risk which determines the required return at which an asset should be valued.

One of the fundamental concepts arising out of the development of Capital Market Theory is that investors will only be compensated for the risk they cannot eliminate through diversification. In other words they are not compensated for all the risks they take on. This notion is extremely important because an investment may be extremely risky in absolute terms but its contribution to a portfolio may mean that it could be considered relatively riskless.

This understanding of the trade off between risk and return has led to the development of asset pricing models as a means of determining equilibrium values. This naturally leads on to the development of the Security Market Line which relates expected return to systematic risk. This relationship is shown in Figure 2. It is a completely different relationship to the Capital Market Line although the two illustrations are similar. The fundamental difference is that, in equilibrium, every asset will lie on the Security Market Line. Any asset which is under-or overpriced will lie either above or below the line. The Security Market Line also refers to individual assets as well as portfolios. The Capital Market Line by contrast only refers to efficient portfolios.

The development of portfolio theory and its evolution into

capital asset pricing has spanned a period from the early 1950's to the present day. This is almost the same period that the investment market in property has been operating. Although the concepts evolved were largely devoted to the analysis and construction of equity portfolios there is nothing in the theory which restricts it to this sector of the market.

The property market is, however, very naive when it comes to taking on board new ideas. Reliable data on property are still extremely difficult to obtain and agents and valuers have evolved a mystique around the whole of the market restricting both the free flow of information and the development of innovative ideas. It is probably for these reasons that there has been very little fundamental research in this country into the application and development of portfolio theory and asset pricing to property investment. There are of course extenuating circumstances. The property booms since the 1960's meant that there was little need for fundamental research. There has also been little need for pension funds to justify their decisions, particularly when the common view was that investment in property was essentially long term.

All that can be said in response to this is that the market is changing. There is considerable pressure from other professions to break down the mystique. Property is no longer regarded as a long term investment. Decisions now have to be justified on more than a hunch and a belief that yields are the only measure of investment potential. It is in these areas that financial theory has a lot to offer the property profession.

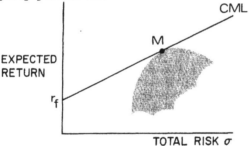

Figure 1 The Capital Market Line (CML)

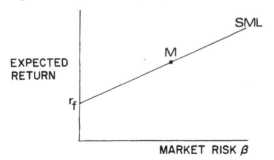

Figure 2 The Security Market Line (SML)

113

PART 2

Part 1 of this chapter showed that it is possible to construct a
portfolio which achieves the best trade off between risk and return.
The algorithm developed by Markowitz enables investors to calculate
the proportion of funds which should be invested in each asset.
Although property is indivisible the principle remains unchanged. In
fact portfolio alglorithms have recently been developed which cope
with problems of asset selection when indivisible assets are
involved. This aspect no longer represents a restriction and if the
unitised property market becomes widely tried then the opportunity
for taking variable positions in property will be considerably
enhanced.

Although the theory for constructing a portfolio is well
understood and models are available for tracing out the efficient
frontier there still remains the problem of choosing a particular
portfolio. Markowitz addressed this by referring to risk
indifference curves. These trace out the attitude of investors to
different levels of risk and return. Figure 3 shows three such
curves related to (a) risk seekers who are willing to take on more
risk on the promise of a high return, (b) risk averters who prefer
less risk to more risk as returns increase and (c) the risk
indifferent investor who is equally happy with a high or low risk
investment. Most investors, however, tend to be risk averse so that
their risk indifferences curves are convex.

Given that a whole family of risk indifference curves can be
drawn for the risk averse investors they can be superimposed on to
the same graph as the opportunity set and where a curve touches the
efficient frontier it designates the optimal portfolio. Given a
family of increasing risk indifference curves an investor will try to
choose the highest curve. In so doing he will maximise this utility
or attitude to risk. This selection procedure is shown in Figure 4.

Given the opportunity to invest at the risk free rate it will be
evident that investors can move on to higher risk indifference
curves. This is shown in Figure 5 which is related to the Capital
Market Line. Portfolio O lies on a higher risk indifference curve
than portfolio P. By varying the proportion of funds between the
market portfolio and the riskless asset it will be seen that the
investor can alter the risk indifference curve on which he lies.

Recall, however, that the Capital Market Line relates solely to
efficient portfolios, i.e. those that only contain systematic or
market risk. A similar interpretation applies if the Security Market
Line is used. Investors can maximise their utility by moving on to
their highest risk indifference curve. If an asset or portfolio is
correctly priced in relation to the market then it will lie on the
Security Market Line. The expected return for the asset or portfolio
will give a net present value of zero when used to discount the cash
flows. If, however, the asset or portfolio is underpriced it will
lie above the Security Market Line. An investor holding such an
asset or portfolio will be able to move on to a higher risk
indifference curve and thus maximise his utility. What is equally
important, however, is that the portfolio or asset will also generate
a positive net present value. It should be evident from this that
selecting portfolios or assets with positive net present values is
entirely consistent with

maximising utility. This is an important point which is probably not widely understood but is nevertheless worth considering. The use of net present value as a decision making criterion will help a fund maximise its value in the market place. At a more general level net present value can be related to share value. If correctly priced the value of a company is merely the present value of its expected future income when discounted at the appropriate risk adjusted return. If its value is in equilibrium then it will have a net present value equal to zero. If the company is undervalued its net present value will be positive. The value of the shares in that company will, therefore, rise by the amount of the net present value. Sensible companies will therefore aim to maximise net present value and will actively seek out those assets which are underpriced.

Looking at the problem of asset selection in this way is important because it is not easy to estimate risk indifference curves. In fact it is probably impossible to establish suitable risk indifference curves for something as nebulous as a financial institution. The solution to this problem is, however, not relevant as the above discussion indicates that as long as assets are chosen which maximise net present value the decision to invest can be made independently of risk class. Having said this investors will generally have a preference for a risk class although these will be determined largely by tax clientele effects.

The Security Market Line identified above is a concept which relates the expected return on an asset to its market risk. In its simplest form it is a graphical representation of the capital asset pricing model. It relates expected return to systematic or market risk. Intuitively it is an easy model to understand and although it cannot be regarded as the ultimate truth concerning risk and return it does provide a very powerful framework for making sound investment decisions.

It is perhaps worth contrasting this approach with the way decisions to invest in property are currently made. The common approach is to base most decisions on an examination of the property yield. This, however, says nothing about net present value or rate of return and will lead to economically correct investment decisions merely by chance. The paradox of this situation is that professional advisors, when recommending a property for purchase, will always claim that the property is underpriced. As this concept is largely misunderstood in the property world it is probable that random selection would achieve returns equivalent to those selected by professional advisors. If the cost of professional advice is taken into consideration then random selection may well prove more effective.

If traditional advice was always correct so that every property was underpriced in the economic sense then evidence of superior decision making would be observable in the market place. Property investment funds, unit trusts and bonds holding large amounts of real property would consistently outperform the market. Empirical analysis shows that this is not the case.

There has in recent years been a move towards discounted cash flow techniques as a means of quantifying the inputs. Valuations resulting from this approach do, however, rely upon accurate forecasts of expected cash flows and the choice of discount rate which acequately compensates for risk. If the figures chosen have no economic basis then the resulting valuations will have little relevance in terms of decision making.

The issues raised in this discussion are important in relation to the efficiency of the property market and the way it uses information. Consider the following scenario.

It is a widely held view that the property market is inefficient. In other words valuations and prices are not an accurate reflection of all available information. Property advisors know that the market is inefficient and are able to select underpriced properties. However if the market is inefficient and properties are underpriced then they must also be underpriced from the point of view of the vendor. This scenario is unreasonable. It is more likely that the market operates in a broadly efficient manner otherwise there would be an excess demand for property.

The question of efficiency also has an impact on depreciation and obsolescence. If the market is operating in a broadly efficient way then values should reflect what is currently known about a property and its future expectations. As we move through time the valuation of a property will change in response to new information. In general yields should increase to reflect a reduction in growth expectations and this trend can be observed in the market. If yields do not take account of depreciation then they must be inadequate processors of information and this will lead the investor into areas of profitable arbitrage.

I find this proposition hard to believe mainly because depreciation is publicly available information and should be incorporated into values. The following quotation from Findlay, Messner and Tarantello[1] summarises the position:

> 'It does seem however, that few people today would seriously assert that real estate is totally different from anything else ... In a reasonably organised market, we would expect some approximation of efficiency and equilibrium pricing to hold. At the very least we would expect the absence of arbitrage opportunities.'

The reason the property market is believed to be inefficient is because of the lack of adqeuate empirical evidence to refute the suggestion coupled with the publication of details of indiviudal property transactions. Property journals frequently contain articles which make oblique references to the inefficiency of the market by publishing details of one or two abnormal transactions. Clearly this is the type of information that property advisors like but it would be unwise to base an investment strategy on such limited evidence.

Research undertaken in both this country and in Canada has shown that the market is efficient at the weak form level. In other words an historic time series of prices says nothing about future price

---

1. Findlay M C, S D Messne and R A Tarantello:1983, Real Estate Portfolio Analysis, Lexington Books

changes so that investors are unable to construct profitable trading strategies merely on the basis of this information. The research in this country was based on valuations conforming to the RICS definition of Open Market Value. The subset of information determining values was very clearly defined. As the subset of information becomes more restricted however, it is possible that the market will become inefficient and thus offer opportunities for profitable arbitrage. An example of this may be the forced sale of a property resulting in a price below its equilibrium market value.

It is under these conditions that the market may well prove to be inefficient. In this example a test of this hypothesis would need to examine a very large sample of forced sales to establish whether actual returns differed significantly from expected returns. Problems of data availability usually prevent this type of analysis being undertaken.

One thing is clear however. The type and quality of information available has a significant influence on valuation. Two valuers can therefore arrive at different values for a property because they have researched the market in different ways. It is this aspect which is fundamental to the decision to invest.

If the market was truly efficient then random selection of properties would result in a portfolio which compensated for its risk. There is enough room, however, for investors to acquire costly information in order to outperform the market. Agents try to fill the information gap by offering specialised knowledge of local markets. As the number of agents increase the market will become more efficient. The introduction of the unitised property market will also contribute to improved efficiency.

The efficient markets hypothesis does not say that it is impossible to outperform the market. It does, however, say that it is impossible to do this consistently. This is where the challenge lies. Professional research departments must first of all recognise what the problems are and then develop models for identifying underpriced opportunities in an economically valid way. The cost of this research must, however, be less than the benefit gained otherwise it has little value. Useful lessons can be drawn from the stock market where considerable research has gone into the development of economically valid equilibrium valuation models. The number of research papers in the equity field is vast. Highly sophisticated models have been developed for estimating asset risk and developing efficient portfolio strategies. All these techniques are aimed at achieving consistent abnormal returns.

By contrast the property literature on valuation models frequently revolves around trivial matters such as the accuracy of valuations calculated in advance or in arrears. New valuation models have been suggested which claim to offer significant advantages over traditional approaches where in reality they are simple variations of the discounted cash flow approach. Considerable emphasis is placed on the mathematical consistency of the models although very little is said about their economic validity. Although mathematical consistency is important this is usually an algebraic matter which is easily sorted out. If, however, the model is a poor representation of the underlying economics then the chances are that it will not help in the decision making process.

This discussion should illustrate that the role of the valuation model is to determine whether a property is under—or overpriced. If an economically valid model is developed the resulting valuations can be compared with the prices at which properties are offered in the market place and a decision can then be made as to whether or not to buy.

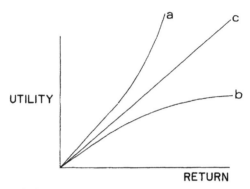

Figure 3 Risk indifference curves

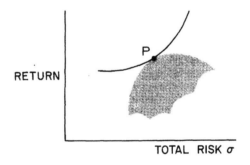

Figure 4 The optimal portfolio (P)

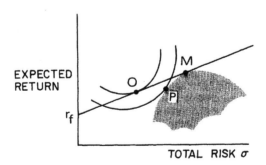

Figure 5 Investment Choice

118

The selection of assets depends on the investor's ability to forecast future return. If investors are using professional advisors then it is the forecasting ability of the professional advisor which is important. Consider therefore the type of scenarios which might exist.

If the investor has perfect forecasting ability he predicts with confidence what the future returns for any property are going to be. His portfolio strategy is simple. All he need do is invest in the single property which is going to offer the highest abnormal return. he should invest as much money as possible in one property.

The investor who has no forecasting ability has an equally easy task. He has no idea what the market is going to do so his best course of action is to hold a portfolio which is very highly diversified and tracks an index. This way he can be sure that he will do no better or worse than the market as a whole.

The difficulty comes in considering the investor who believes he has some forecasting ability. This would imply that a policy of diversification should be followed but the investor should also speculate on those properties which are believed to be underpriced.

In the property world the likely scenario is that investors believe that they have perfect forecasting ability but behave as though they have none. In other words every property they buy is underpriced but they generally follow a policy of diversification in order to track an index!

This behaviour raises the issue of the effectiveness of diversification of property portfolios. Again the market has a naive view of this aspect. It is frequently the case that in professional reports given to clients there will be a statement to the effect that 'the property portfolio is well diversified across all sectors of the market'. The implication is that the investor need not worry too much about his diversification policy as all the risk in his portfolio has been diversified away.

Unfortunately this proposition is untrue. Empirical evidence suggests a completely different picture. It will be recalled from Part 1 that the total risk on an investment is made up of two components: systematic and specific risk. The process of diversification endeavours to eliminate or at least reduce the specific risk to a very small amount. If it could be completely eliminated then all that would remain would be systematic risk and the returns would then track an index. It is possible to test the effectiveness of this by examining a large sample of properties to see how much specific risk remains in a portfolio as more properties are added. Assuming equally weighted properties the following table summarises what is happening.

Returns are expressed in % per annum

Number of Props	Total Risk	Market Risk	Specific Risk
1	15.73	4.67	15.02
10	6.08	4.26	4.34
20	5.55	4.51	3.24
30	5.37	4.63	2.72
40	5.28	4.71	2.39
50	5.22	4.75	2.17
100	5.12	4.87	1.57
200	5.06	4.93	1.13
1000	5.02	4.99	0.51

The second and fourth columns show that both the total risk and specific risk figures reduce as the number of properties in the portfolio increase. The market risk, as expected, hardly changes. The specific risk reduces very rapidly but then tails off slowly. It is this column of figures which shows how much risk remains in the portfolio after diversifying. A portfolio holding 20 properties, for example, can still have about 3% per annum specific risk. This can be interpreted as follows. The market risk will determine the expected return on a portfolio. Assume that the market risk given above is equivalent to a beta of 1.0 and that the risk free rate of return is 9%. Assume also that the security market line is upward sloping and that the market premium is 2%. This gives us enough information to calculate the expected return for our 20-property portfolio. Using the capital asset pricing model it's expected return would be as follows:

$$E(r_p) = r_f + \beta_p (r_m - r_f)$$

$$11.0 = 9.0 + 1.0(2.0)$$

Although the portfolio has an expected return of 11.0% it also carries 3.0% per annum specific risk. This is expressed in terms of a standard deviation so that we can say that there will be approximately a 66% chance that the portfolio return will lie in the range of 11.0% ± 3.0%, i.e. from 8% to 14%. If we take three standard deviations we can be 99% certain that the portfolio return will lie in the region of 11.0% ± 9.0% or from 2% to 20%. The amount of specific risk can mean the difference between over or under-performing the market.

This is very important because most portfolios tend to hold less than 30 properties so that the amount of specific risk they have is very high. The immediate implication is that the performance of a portfolio will be determined largely by the individual properties rather than any ability to diversify. It also implies that professional advice concerning diversification is wrong. It also points to the need to improve the quality of advice and research required to select underpriced properties. Unfortunately, this is an area which is sadly lacking in most professional firms.

A further implication of examining risk reduction potential is that there appears to be little difference between sectors. A portfolio consisting entirely of office properties will have about the same level of risk reduction as one which consists entirely of retail or industrial properties. Diversification across sectors does not therefore appear to be as worthwhile as is generally imagined. What is worthwhile, however, is diversifying within a sector which has the greatest upside potential. Over the last few years this has been the retail sector but there is no guarantee that this will always hold.

The desire to diversify across sectors is probably influenced by the long holding periods associated with property. A portfolio diversified across all sectors which is not actively traded may well gain if the upside potential of individual sectors changes.

In order to eliminate a large proportion of specific risk it will be seen that a portfolio must hold a great many properties. However, even a portfolio with 1000 properties still has about 0.5% specific risk per annum which could be diversified away.

This has implications in terms of the construction of property indices as one with only a small number of properties is unlikely to be tackling the market accurately.

Because of the problems outlined above the rationale for diversification must be questioned. If it were possible to construct a portfolio which tracked an index the immediate implication would be that it could not outperform the index. The only way that this would be possible would be by reducing the level of diversification. Although this is difficult with an equity portfolio it is straightforward with a property portfolio. In fact it is probably true to say that most, if not all, property portfolios are poorly diversified.

The performance of portfolios therefore depends on the ability of investors or their advisors to select properties which are underpriced or manage a portfolio in such a way that its net present value is maximised. Although it has been recognised in recent years that property portfolios should be actively managed I know of no fund which uses net present value as the principal criterion to be maximised.

The only way that superior performance can be detected is for the portfolio to be analysed using a performance measurement system.

There are a number of these available on the market but they all suffer from the following three main problems.

(i)   The reference index used usually has a high level of serial correlation in the returns. This is induced by averaging the values round the publication date of the index number. Property indices are therefore moving averages. When used in performance measurement it is difficult to attribute abnormal performance to superior investment skill or the fact that the index number is not a true reflection of the underlying market. Serial correlation is a major problem and can be very high for property indices. A byproduct of high serial correlation is that it is possible to forecast with a high degree of accuracy what the return will be in the next period. Given this information it is then possible to construct a trading strategy which will capitalise on that information. The reason this does not happen in practice is because the high levels of correlation are caused by statistical problems in the way the index is calculated.

(ii) The risk of the portfolio under consideration is assumed to be exactly the same as the market. Abnormal performance may therefore be due to differences in risk class and not management skill.

(iii) The index used almost certainly carries some specific risk which could be diversified away. The index return may not therefore be the true return.

The combination of these factors means that commercial property performance measurement systems in reality say very little about real performance and it is doubtful whether the output is at all useful as a means of guiding investment strategy. The position is unlikely to change unless there is a greater emphasis on justifying investment decisions. There is some evidence to suggest that this is beginning to happen although the majority of professional advisors know practically nothing about modern finance.

The attached note describes part of a new system of performance measures developed by Property Research Services which recognises the problems in conventional systems and solves them in a way which is economically valid. The system has been developed out of several years of empirical research and forms part of a planning tool aimed at developing and monitoring a strategy for property portfolios. Much of the information produced by this system is new but gives a better indication of the way the portfolio is performing. Included within the analysis are the following:

(a)   risk estimates for each property and the portfolio
(b)   estimates of abnormal return
(c)   abnormal returns converted to net present values
(d)   estimates of risk adjusted required returns
(e)   estimates of expected growth
(f)   an analysis of abnormal return
(g)   an estimate of portfolio diversification
(h)   an estimate of the portfolio specific risk
(i)   an estimate of the reduction in risk achieved
(j)   estimates of upside potential and downside risk

As none of this information is currently available their use in the analysis and interpretation of property portfolios represents a major advance on conventional systems.

The analysis contained in this section is based on the principles described earlier and attempts to overcome many of the problems associated with conventional performance measurement systems.

It should be stressed that the system described here is not part of a valuation package. The inputs for the analysis must be derived from independent valuations. What is being offered is a strategic approach to analysis aimed at maximising the market value of investment portfolios using methods that are economically valid.

The system is therefore based on the valuations prepared by professional valuers. It uses available information about the portfolio and the market to estimate the risk class[1] of each property and to assess the net present value of the total portfolio. In addition it analyses how well diversified is the portfolio and quantifies the amount of variation likely in the returns from year to year.

After adjusting for risk the abnormal returns for each property are quantified over the period of analysis. They are further analysed in order to determine whether there is any significant bias in the valuations.

The analysis also proceeds to estimate the equated yield of each property together with its expected growth. This information is not available elsewhere and is potentially useful in providing a data base of comparable evidence. By utilising the information contained in the analysis a knowledge based report is prepared which explains the performance of the portfolio in plain English.

One of the features of the analysis is the ability to restructure the portfolio in order to examine the effect of changing valuations or lease structure on the expectations and diversification of the portfolio.

The measures of risk shown in the following analysis can best be interpreted as the factor which should be applied to the average market risk premium in order to arrive at a target return for the property. The target returns are shown as equated yields. This figure is intended to represent the expected return on each property assuming they were held for long periods. It is also assumed that in order to estimate these figures investors on average expect a premium of 2% from a very large portfolio diversified across each sector. Thus if the risk free rate of return is taken as, say, 10% and the risk factor for a particular property is assessed at 0.5 then the target return (i.e. equated yield) for the property, assuming an average market premium of 2%, would be calculated as follows:

Target Return = 10% + 0.5(2%) = 11%

This figure is of course based upon the valuation prepared at the beginning of the year by the valuer. It is therefore the target return implied by the valuation. Given the target return it is then possible to estimate the expected growth rate implied by the valuation. Other than using subjective judgement these two valuable pieces of information are not available through any other analysis system.

---

1. The method used for assessing risk is covered in: Brown G R, 'A Certainty Equivalent Expectations Model for Estimation of the Systematic Risk of Property Investments', Journal of Valuation (Vol.6 No.1 pp17-41).

MARKET DATA
REFERENCE INDEX:

Richard Ellis Property Market Indicators: 1984-85

% WT	B Risk	Sector	Equated Yield	Expected Growth	Equivalent Yield
50.64	1.05	Office	13.03	7.29	7.12
28.21	1.42	Retail	13.77	8.95	5.61
21.15	0.32	Industrial	11.57	1.68	10.19
	1.00		12.93	6.57	7.34

Risk free return	10.93%
Market return	7.72%
Capital growth	2.32%
Income yield	5.40%
Equivalent yield	7.34%
Rental value yield	7.72%

XYZ PORTFOLIO SUMMARY
(1984-85)

% WT	B Risk	Sector	Net Present Value £	Equated Yield	Expected Growth	Equivalent Yield
48.00	0.85	Office	109,793	12.65	5.06	8.06
20.00	1.70	Retail	524,938	14.60	10.60	4.75
32.00	0.68	Industrial	-2,052,750	12.13	4.28	8.47
	0.97		-1,418,018	12.87	5.92	7.53

Number of properties	41
Equal equivalent	22
Income yield	8.40%
Actual return	5.38%
Expected return	7.81%
Abnormal return	-2.43%

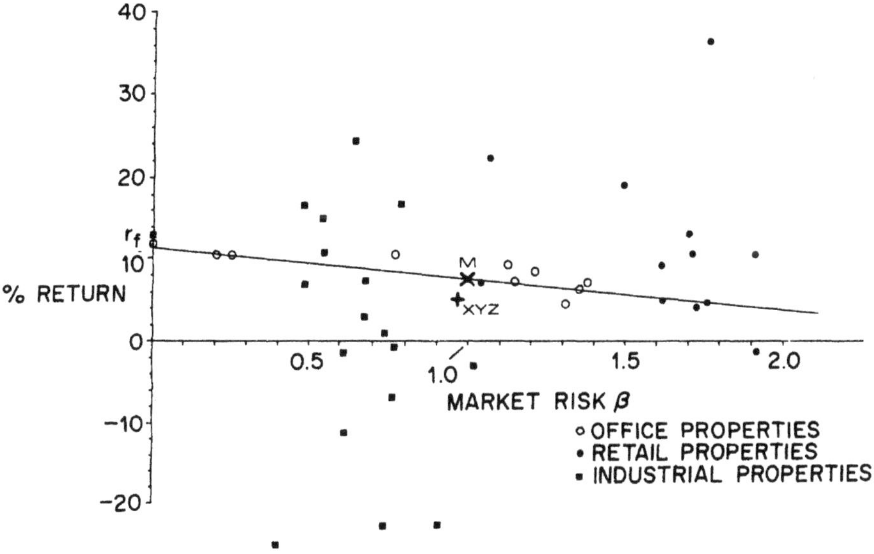

Figure 6

125

ANALYSIS OF ABNORMAL RETURN

Abnormal Return	−2.43%

This can be split as follows:

Effect of diversification	−0.73%
Effect of selection	−1.70%

ANALYSIS OF DIVERSIFICATION

Market effects explain the following percentage of portfolio returns	65.55%
Factors specific to the portfolio which have not been diversified away explain the remaining variation	34.45%
Because the portfolio is poorly diversified actual returns will differ from expected returns by at least the following amount	3.05%
The policy of diversification has managed to achieve the following level of risk reduction over the average property risk within the portfolio	62.62%

ANALAYSIS DATE   1984-85
PORTFOLIO NAME   XYZ PROPERTY FUND

Assuming upward only rent reviews

Prop	Risk	Equated Yield	Expected Growth	Equivalent Yield
0:001	1.12	13.17	6.57	7.32
0:002	1.30	13.53	8.21	6.37
0:003	0.26	11.45	1.67	10.05
0:004	0.00	10.93	-6.61	15.40
0:005	1.14	13.21	7.31	6.82
0:006	0.20	11.33	1.28	10.26
0:007	1.38	13.69	8.58	5.90
0:008	1.38	12.47	4.88	8.23
0:009	1.20	13.33	7.50	6.58
0:010	1.35	13.63	8.40	6.01
R:011	1.76	14.45	10.94	4.58
R:012	1.76	14.45	10.97	4.57
R:013	1.91	14.75	11.82	3.97
R:014	1/91	14.75	11.75	3.66
R:015	1.70	14.33	10.48	4.59
R:016	1.07	13.07	6.71	7.08
R:017	1.74	14.41	10.74	4.37
R:018	1.62	14.17	10.02	4.91
R:019	1.04	13.01	6.53	7.20
R:020	1.70	14.33	10.50	4.57
R:021	1.49	13.91	9.25	5.42
R:022	1.63	14.19	10.10	4.88
I:023	0.60	12.13	3.90	9.00
I:024	0.00	10.93	-1.29	12.32
I:025	0.68	12.29	4.40	8.72
I:026	0.74	12.41	4.80	8.49
I:027	0.67	12.27	4.34	8.74
I:028	0.76	12.45	4.84	8.30
I:029	0.91	12.75	5.73	7.70
I:030	0.76	12.45	4.80	8.30
I:031	0.00	10.93	-0.51	11.35
I:032	0.49	11.91	3.12	9.26
I:033	0.64	12.21	4.10	8.72
I:034	0.76	12.45	4.79	8.29
I:035	0.78	12.49	4.92	8.21
I:036	0.46	11.85	2.94	9.39
I:037	0.54	12.01	3.45	9.05
I:038	0.40	11.73	2.53	9.55
I:039	1.04	13.01	6.51	7.20
I:040	0.61	12.15	3.96	8.94
I:041	0.54	12.01	3.53	9.18
Average	0.96	12.84	5.79	7.64

Assuming upward only reviews

Prop Ref	% Prop Weight	B Risk	Expected % Return	Property % Return	Abnormal % Return	Net Pres Value (£)
0:001	4.62	1.12	7.33	9.48	2.15	56,470.39
0:002	2.43	1.30	6.74	4.60	-2.14	-38,773.57
0:003	10.37	0.26	10.10	10.60	0.50	33,243.81
0:004	1.95	0.00	10.93	10.93	0.00	0.00
0:005	15.99	1.14	7.27	7.40	0.13	14,928.95
0:006	3.78	0.20	10.29	10.36	0.07	1,713.19
0:007	2.18	1.38	6.51	7.40	0.89	12,413.27
0:008	0.78	0.77	8.45	10.93	2.48	11,791.80
0:009	2.71	1.20	7.07	8.12	1.05	18,076.32
0:010	3.22	1.35	6.60	6.60	-0.00	-70.72
R:011	0.68	1.76	5.29	36.20	30.91	157,294.40
R:012	0.82	1.76	5.27	4.80	-0.47	-2,862.27
R:013	1.39	1.91	4.81	10.60	5.79	61,719.61
R:014	6.87	1.91	4.78	-1.80	-6.58	-343,815.40
R:015	0.63	1.70	5.49	10.90	5.41	25,452.04
R:016	2.65	1.07	7.49	23.40	15.91	322,752.20
R:017	1.71	1.74	5.33	4.30	-1.03	-13,415.40
R:018	2.14	1.62	5.73	5.60	-0.13	-2,092.81
R:019	0.22	1.04	7.58	7.50	-0.08	-132.81
R:020	1.08	1.70	5.47	13.80	8.33	66,647.93
R:021	0.96	1.49	6.14	19.74	13.60	218,864.80
R:022	1.24	1.63	5.69	9.40	3.71	34,525.61
I:023	2.21	0.60	9.00	-1.60	-10.30	-162,662.80
I:024	0.70	0.00	10.93	33.30	22.37	106,673.00
I:025	2.86	0.68	8.75	2.80	-5.95	-117,939.40
I:026	2.60	0.74	8.54	1.20	-7.34	-132.704.40
I:027	1.09	0.67	8.78	7.50	-1.28	-9.590.60
I:028	2.54	0.76	8.48	-22.50	-30.98	-540,172.30
I:029	0.99	0.91	8.01	-22.40	-30.41	-21,518.46
I:030	5.02	0.76	8.50	-0.10	-8.60	-302,468.20
I:031	0.06	0.00	10.93	12.10	1.17	438.04
I:032	1.02	0.49	9.36	16.70	7.34	51,583.99
I:033	0.91	0.64	8.86	24.10	15.24	95,885.51
I:034	2.63	0.76	8.50	-6.50	-15.00	-275,682.80
I:035	0.76	0.78	8.43	16.50	8.07	35,435.16
I:036	0.48	0.46	9.46	6.80	-2.66	-8,909.49
I:037	1.74	0.54	9.19	14.60	5.41	67,701.60
I:038	1.97	0.40	9.66	-24.80	-34.46	-509,711.60
I:039	2.48	1.04	7.60	-3.50	-11.10	-196,163.80
I:040	1.02	0.61	8.97	-10.80	-19.77	-138,220.60
I:041	0.48	0.54	9.19	10.80	1.61	5,277.57

Portfolio Risk	0.97 (value wtd)	Portfolio NPV	-£1,418,018
	0.96 (value wtd)	Portfolio Value	£82,218,150

128

Analysis Period 1984-85

Portfolio Value

On the basis of the risk adjusted discount rates estimated for each property over the period 1984-85 the portfolio had a negative net present value of -£1,418,018. Although the portfolio had a positive return of 5.38% this was insufficient to compensate for the risk taken and will result in a fall in the share value of the investing fund equal to the net present value.

This statement relates solely to the analysis of the standing investments and ignores other factors which may influence the net present value. For example, proceeds from disposals and developments in progress have been ignored together with any other investment activities taken on by the fund. In addition the analysis has been carried out on a gross of tax basis so that any benefit, derived from tax shelters or investment incentives, where applicable, have been ignored. The net present value of these other activities would have to exceed £1,418,018 in order to show an increase in the share value of the fund.

Market Conditions

Conditions in the property market over the analysis period were poor resulting in a market return of 7.72% which was below the return which could have been achieved by investing in a riskless asset with a one yearly maturity, viz 10.93%.

The property market has shown a negative premium over the risk free return for the last five years. Despite these poor market conditions 22 of the 41 properties were able to achieve positive abnormal returns during 1984-85.

Market conditions such as these will favour properties with low expected risk. These are currently to be found in the industrial sector. The actual performance of individual sectors will, however, depend to a great extent on the upside potential and downside risk of each sector. This aspect will be discussed later.

Diversification

Despite the fact that the portfolio holds 41 properties in economic terms it is still regarded as being poorly diversified. On the basis of the risk levels computed for each property it is estimated that the property market explains about 65.55% of the variation in the returns of the portfolio. The remaining 34.45% can be attributed to factors specific to the individual properties. Although the risk which remains in the portfolio could be diversified away it would require the addition of many thousands of properties. In practical terms the risk which has not been diversified away amounts to 3.05% per annum. As long as there is no change in the risk class of the portfolio the actual return achieved in any year will differ from its expected return by at least 3.05%.

In terms of risk the diversification policy has managed to reduce the average property risk in the portfolio by 62.62%. This is close to the limit which can be achieved so that it is unlikely that adding more properties will significantly reduce this figure. Further properties should only be added if they increase the net present value of the portfolio. Analysis of candidate properties should be made against their expected return after adjusting for risk.

Examination of the distribution of funds reveals that over 25% of the portfolio by value is contained in two properties. However, the difference between the value weighted and equally weighted risk measures for the total portfolio is so small that it indicates that the effect of the large properties has been diversified away. In addition the abnormal returns for the two properties (0:003 and 0:005) are close to zero indicating that the valuations placed on the properties have compensated for their risk. The net present values of the two properties are also small relative to their capital values.

The market risk class of the portfolio at 0.97 is very close to 1.00. If this does not change over time then the long term expected return on the portfolio will be close to the market. However, if there is no change in portfolio composition then the expected return on the portfolio will decline relative to the market unless some of the properties become marginal. The distribution of funds and the risk classes of each sector are broadly in line with the market although there is a heavier concentration in the industrial sector. The evidence, suggests however that the fund may well be following a policy of trying to track the market. Although this is impossible in terms of periodic rates of return it is quite possible to achieve in terms of expectations.

Although there may be some errors in valuation which would affect the estimation of abnormal returns and net present values for each property such errors would only be significant at the portfolio level if there was significant bias in the valuations. This, however, does not appear to be the case as the abnormal returns appear to be normally distributed. Errors in valuation have therefore been diversified away.

The abnormal return for the portfolio has been estimated as -2.43%. This can be split into two components, part of which can be attributed to the diversification or policy decision pursued by the fund and the remainder which can be attributed to selection ability. Thus over the analysis period the porfolio achieved a return which was 2.43% less than required to compensate for its risk class. Of this figure 0.73% could be attributed to the diversification policy and the remaining 1.70% to the ability to select properties. This finding is consistent with a portfolio which is poorly diversified. Most of the abnormal return could be attributed to factors specific to the individual properties. Improving performance can only be achieved by selecting those properties which have positive NPV's.

Performance of Sectors

Examination of the net present value of the properties shows that the industrial sector has the highest proportion of negative values. In total the NPV of the industrial sector is -£2,052,750. This represents 7.91% of the capital value of all the industrial properties. The industrial sector is currently showing a lot of downside risk which is reflected in the abnormal returns. Although some of the properties showed positive abnormal returns their effect on the NPV was insufficient to create positive value.

By contrast the retail sector is showing a lot of upside potential which has created high abnormal returns over the last few years. Although this is reflected in the portfolio the abnormal returns for the retail properties are not consistently positive. The net present value of this sector is £524,938 which represents 3.13% of the total value of the retail properties.

The office sector tends to dominate the property market in terms of value and as a result has had smaller abnormal returns than the other sectors. A similar picture emerges with the portfolio. The abnormal returns are small relative to the other sectors and this is also reflected in the smaller net present values. The NPV for the office sector is £109,793 which represents 0.28% of the total value of the office properties. The portfolio has considerable downside risk, most of which can be attributed to the industrial sector. The downside risk can be quantified by summing the net present values which are greater or less than zero.

$$
\begin{array}{ll}
\text{Total NPV's} > 0 = & \text{£1,398,889} \\
\text{Total NPV's} < 0 = & \text{-£2,816,907} \\
\hline
\text{Difference} \quad\quad = & \text{-£1,418,018}
\end{array}
$$

The difference between the two figures represents, in this case, the downside risk of the portfolio.

The upside potential (UP) and downside risk (DR) for each sector can be summarised in a similar manner.

Office Sector

$$
\begin{array}{lll}
\text{Total NPV's} > 0 & = \text{£148,637} & \text{(UP)} \\
\text{Total NPV's} < 0 & = \text{-£38,844} & \text{(DR)} \\
\hline
\text{Difference} & = \text{£109,793}
\end{array}
$$

Retail Sector

$$
\begin{array}{lll}
\text{Total NPV's} > 0 & = \text{£887,257} & \text{(UP)} \\
\text{Total NPV's} < 0 & = \text{-£362,318} & \text{(DR)} \\
\hline
\text{Difference} & = \text{£524,939}
\end{array}
$$

Industrial Sector

$$
\begin{array}{lll}
\text{Total NPV's} > 0 & = \text{£362,995} & \text{(UP)} \\
\text{Total NPV's} < 0 & = \text{-£2,415,745} & \text{(DR)} \\
\hline
\text{Difference} & = \text{-£2,052,750}
\end{array}
$$

Portfolio strategy

In order to improve the performance of the portfolio a more aggressive approach is needed in terms of the buy/sell policy. An active management approach is consistent with a portfolio which is poorly diversified.

Examination of the acquisition dates indicates that there have been a number of properties bought within the last 5 years. Most of these, however, have been within the industrial sector. Many of the other properties were acquired between 15 and 17 years ago. The age structure of the portfolio is therefore consistent with an inactively managed portfolio.

Using a constant 2.00% margin against each level of risk those properties which have an abnormal return lower than this figure should be considered for disposal. The 2.00% margin is intended to take account of potential error in the index returns due to the fact that it contains some residual risk. Figure 6 (p.125) plots each of the property returns against their level of risk and compares them with the risk/return relationship existing during the period 1984-85. The 2.00% margin is also shown from which can be identified those properties which ought to be considered for disposal. They are as follows:

Prop	% Return	% Abnormal % Return
O:002	4.60	−2.14
R:014	−1.80	−6.58
I:023	−1.60	−10.60
I:025	2.80	−5.95
I:026	1.20	−7.34
I:028	−22.50	−30.98
I:029	−22.40	−30.41
I:030	−0.10	−8.60
I:034	−6.50	−15.00
I:036	6.80	−2.66
I:038	−24.80	−34.46
I:039	−3.50	−11.10
I:40	−10.80	−19.77

O = Office Property; R = Retail Property; I = Industrial Property

To be precise about which properties to sell would require collecting data on abnormal returns over periods of between three and five years. Those properties with consistent negative abnormal returns would therefore be identified as suitable candidates for disposal. Although an individual property may have a significant high or low abnormal return in any year this may be as a result of chance events. Collecting data over a cycle of three to five years should eliminate this possibility and enable the opportunity for sale and disposal costs to be recouped. Alternative methods of actively managing the portfolio should also be considered. These could include restructuring the leases to bring them in line with modern standards and undertaking selective refurbishment. Funds should only be committed to refurbishment if the resulting work creates an increase in net present value.

CHAPTER 8

ASSET PRICING MODELS AND PROPERTY AS A LONG TERM INVESTMENT: THE
CONTRIBUTION OF DURATION

1 Introduction

The analysis of property investment has over the past decade taken a
predictable but significant shift. Increasingly research
internationally is becoming influenced by the concepts developed in
the financial economics literature. Real estate texts now mention
portfolio implications as a matter of course and investors are keenly
aware that the decision to invest in property is a decision not to
invest in alternative assets such as ordinary shares or bonds. This
process is being accelerated by the competition amongst firms to
seize the profitable business of advising institutional investors on
property investment. It is no longer novel to read of property
specialists hired by predominantly financial institutions to offer
advice and comment on the property market. There are thus two
related trends which are evident, particularly in the US; finance
academics are publishing more about the reported returns from real
estate and financial institutions are taking a more enthusiastic
interest in property investment.
   The objective of this chapter is to discuss how the developments
in the finance literature can explain some key issues underlying
these trends. In particular, I aim to answer two questions,

   (i) What does property offer the long term investor?
   (ii) How risky is property as a long term investment?

   Obviously the questions are not independent but the answers
involve quite different strands of finance and investment theory.
   Both strands however will be increasingly important to researchers
and professional investors alike over the next decade.

2 What does Property offer the Long Term Investor?

To institutional investors the benefit offered by property investment
is that of owning an asset which combines the security of a corporate
bond with the potential profitability of an equity. In cash flow
terms, the income from property will be secure from one rent review
period to the next, although the capital value will be subject to (a)
changes arising from interest rate movements and (b) changes arising
out of changed expectations in the rental growth rate. Even these
changes can be assigned bond-like or equity-like factors. The
central point therefore is that in order to understand the factors
generating returns in property, we shall have to refer to the factors
which have been found to be useful in analysing both equities and
Bonds. In this section I shall restrict my discussion to the
implications of bond (portfolio) research.

The bond literature has a long history but in terms of the conventional investments text contains many concepts which will be familiar to property researchers. The traditional explanation as offered, for example, in earlier editions of Fischer and Jordan:1987 starts by explaining the calculations of various component returns, income yield, redemption yield or internal rate of return. There is then a short discussion on different aspects of risk such as the risk of default, the risk of interest rate changes and the risk of inflation. This will lead on to a discussion of the relationship between the yield on bonds and the maturity - the term structure of interest rates. It is almost irresistible for writers in this genre to illustrate the term structure by a diagram of the yield curves for government securities. The only restraint on the enthusiasm of authors in this respect is that for the last four or five years the yield curve has been sloping gently downwards - long term rates have tended to be no higher and sometimes significantly lower than short term rates. This recent pattern is unfortunate because of the illustrations which are invariably supported by a discussion of the theories which 'explain' why the long term rates are 'usually' higher than short term rates. Explanations involve expectations or a clientele based demand for bonds of different maturity or a risk averse attitude towards tying up funds in a long term investment subject to interest rate risk. Finally the section finishes by referring to taxation considerations and in the case of US texts especially, to bond rating services and the assessment of 'quality' (i.e. estimates of the probability of default).

The discussion on bond valuation is similar to the way in which property investment is considered in a number of respects. Emphasis is laid on valuation-by-formula, the institutional differences of the type of bond (property), the influence of taxation and most importantly of all, the viewing of a bond (property) in terms of a single investment decision in which the investment is held in perpetuity or until the asset matures, expires or is redeemed. In the case of property there is an additional reason for the ritual procedures which surround valuation; valuations are required for specific purposes which strictly define the terms in which the valuation is framed. Thus it would be implausible to arrive at a valuation for mortgage purposes by the same process as that involved if there was a large degree of discretion in deciding when the property was to be sold, i.e. for trading in the investment market. The key similarity when comparing the investment valuations of both property and bonds is the lack of any explicit recognition of port- folios. 'The property is cheap and should be bought.' 'The bond is offering a lower yield and should be sold'. These are single-asset considerations that are miles away from the equity-influenced models of portfolio construction which have been discussed by Brown:1984 or Fraser:1986. Where then do we find the references to the assessment of bonds (or property) in terms of their value as one part of the portfolio? The answer is historically in the actuarial literature (Redington:1952) but more recently in the US financial research on pension fund management (see for example Leibowitz:1986).

The portfolio investment decision in this approach is seen as a joint analysis of the fund's assets and liabilities. The liabilities are the future benefits which will be paid out to the fund beneficiaries. If the contributions rise in money terms, the future liabilities will also increase. If external factors such as the long term rate of interest change, the present value of the liabilities will change appropriately. One important factor which is identified as causing the present value of liabilities to alter is the rate of interest. Leibowitz and the other writers in this area thus emphasise the sensitivity of the valuation to changes in the rate of interest. This is encapsulated in the measurement of the fund's assets and liabilities in terms of the 'duration'.

Duration can be defined as the elasticity of the asset price or value, to changes in the rate of interest. Intuitively it can also be thought of as a 'half-life' of the asset. (See Boulding:1936 for a very early recognition of this aspect.) If a pension fund manager wants to minimise his/her risk, the duration of the assets should be matched to the duration of the liabilities. If the duration of the assets is lower than the duration of the liabilities, a fall in the rate of interest will increase the value of the assets by less than the increase in the value of the fund's liabilities; the fund would therefore be less successful. Any mismatch between the assets and liabilities can therefore be seen as speculating on the movement of rates of interest. Fund managers may be successful in their investment decisions and the fund may be very profitable but on balance the matching of the assets and liabilities is the minimum risk strategy.

In this framework, we can see that we have an additional dimension of risk introduced into the valuation equation. Low duration assets are less volatile than high duration assets but if they do not match the fund's liabilities they can be seen as adding risk to the portfolio. Similarly it is not enough to calculate the redemption yield (or internal rate of return) on an asset and conclude that it is cheap or expensive. Instead investors have to value an asset in terms of the needs of their liabilities. Thus by incorporating the concept of duration we have introduced portfolio considerations into bond valuation.

There are two issues which arise from this. How do we estimate the duration of assets (and property in particular) and how does it explain the investment behaviour of investors?

3 Estimation of Duration

Duration is defined as the average time to the receipt of each cash flow, weighted by the present value of the cash flow. In a simple example, if we had cash flows of 10 (in one year's time) and 110 (in two years' time), with an interest rate of 10%, the present value would be 100. The present value of the first payment of $10/1.1=9.09$ whilst the present value of the final payment is $110/1.21=90.91$. The duration is thus given by

$$D = (9.09 \times 1 + 90.91 \times 2)/100 = 1.91 \text{ years.}$$

It will be obvious that the duration is always less than the maturity of the investment. With a bond in which the final payment is a large part of the cash flow, the duration may be longer than half the maturity but for an investment which pays equal cash flows (such as a mortgage) the duration will be less than half the maturity.

In fact for the bond there is a closed form solution of the duration (Chua:1984):

$$D = \frac{C.\left[\dfrac{(1+y)^{M-1} - (1+y) - y^M}{y^2(1+y)^M}\right] + \dfrac{(F)(M)}{(1+y)^M}}{B}$$

Where M = maturity of the bond
      C = coupon payment per period
      F = nominal value
      y = redemption yield
      B = present value of the bond (discounted at y)

I have calculated the duration of some bonds varying by coupon and by maturity (see Table 1). It will be immediately obvious that even bonds with long maturities have relatively short durations; low coupon bonds have greater durations but in none of the examples estimated is the duration greater than 15 years. In fact we can use the elasticity measure of duration to estimate some upper limits. The duration can be estimated by the expression;

$$D = - \frac{dV}{dr} \cdot \frac{(1 + r)}{V}$$

where dV/dr = the (partial) differential of V, w.r.t. r
      V = the value of the asset
      r = the discount rate

Applying this to an irredeemable low coupon bond, say 2.5% bond yielding 10%, i.e. a price of 25, we can make use of a simplified formula. The duration for all irredeemable fixed interest bonds is given by D = (1+r)/r. Thus the duration of 2.5% bond yield 10% is 11 years.

137

Table 1   Estimated durations for representative bonds

Maturity	Coupon	Value	Redemption Yield	Duration
5	10	100	10.00	4.17
5	5	81	10.01	4.49
5	2.5	71.6	9.99	4.71
10	10	100	10.00	6.76
10	5	73.1	10.01	7.78
10	2.5	53.9	10.00	8.50
15	10	100	10.00	8.37
15	5	64.3	10.01	9.84
15	2.5	42.9	10.01	11.14
25	10	100	10.00	9.98
25	5	54.6	10.00	11.25
25	2.5	31.9	10.01	13.24

The liabilities of pension funds are both long term and have long
durations. There is a problem in calculating the duration of the
liabilities because as Ambachtsheer:1987 has pointed out, the pension
fund liabilities can be viewed either in strict legal (actuarial)
terms or as a going concern in which the pension plan sponsors try to
forecast the outgoings in likely scenarios, e.g. with higher levels
of inflation. Since most pension funds have raised the level of
payments made to beneficiaries by a greater amount than the level
strictly due, the liability is likely to be both greater and more
sensitive than that defined by the application of a strict actuarial
calculation.

This argument suggests the need for assets which have durations
than bonds. Equities have been justified on the basis that they have
long durations. If we use the dividend growth model of equity
valuation,

$$V = D_o/(r-g)$$

Where D_o = dividend payable at end of year
      r = discount rate
      g = growth rate of dividends

Applying the elasticity model of duration, we find

$$D = - (1+r)/(r-g)$$

It is comparatively easy to find equities with durations which
might reasonably be expected to be greater than 15 years by this
estimate. It therefore follows that equities can be used in a
portfolio to reduce risk rather than to increase it, through the
characteristics of long duration and the viewpoint of duration

matching.

Property, as mentioned above, can be seen as a mix of bond-type and equity-type characteristics. In duration terms, this is also true because the cash flows from property tend to rise nominally over time (like dividends) and yet because the appropriate discount rate is lower than that of equities, the difference between the discount rate and the growth rate is low. (In other words the duration of property will be greater for a given yield than the duration of equities.) Judged by the criterion of asset/liability matching, therefore, property is of enormous importance to long term funds since it provides an efficient source of long duration matching.

What is the appropriate measure of duration if we are considering the conventinal n-year reviewed freehold? Given the formula for a DCF-based valuation, it is tedious but not mathematically complex to derive the duration of the property by differentiating the formula with respect to r. The results of this exercise are shown in Appendix 1. Table 2 presents illustrative estimates of the duration of various property interests, ranging from a short leasehold to (approximately) freehold. It will be seen that the durations of the property investment are longer than the equivalent bond for all cases where positive growth is expected.

Table 2   Duration of property investments (3 year rent reviews)

Maturity	Growth Rate (%)	Yield (%)	Duration (Years)
5	0	27.7	2.77
10	0	17.7	4.58
	2.5	16.9	4.73
	5.0	16.1	4.87
	7.5	15.3	5.02
	10.0	14.5	5.16
15	0	14.7	5.98
	2.5	13.5	6.34
	5.0	12.3	6.72
	7.5	11.2	7.10
	10.0	10.01	7.47
20	0	13.4	7.02
	2.5	12.0	7.65
	5.0	10.5	8.31
	7.5	9.2	9.01
	10.0	7.9	9.72
100	0	12.0	9.33
	2.5	9.3	11.73
	5.0	7.6	15.75
	7.5	5.2	23.06
	10.0	2.9	36.05

4   How does Duration explain Behaviour?

The main problem with both the above derivation and the derivation
for the duration of equity is that in practice if interest rates
change, the values of the assets do not behave consistently with the
formulae.   Theoretically this will occur because the growth rate is
not independent of the interest rate.   In simple terms, if growth
rates are highly correlated with the interest rate, the value of the
asset will be insensitive to interest rate changes.   In the case of
equities, it is an empirical result that equity values behave as
though they have a much shorter duration than that implied by the
formula above.   In the case of property also much work remains to be
done before the empirical issues can be determined.   It therefore
remains a theoretical and empirical issue which will be increasingly
important as funds re-enter the institutional investment market.

Table 3   Sensitivity of Duration to the Correlation Between ERV
Growth and Interest Rates

Maturity	ERV Growth Rate	Sensitivity of Growth to Interest Rates		
		0	0.5	1
5	0	2.77	4.02	3.12
	2.5	4.73	4.10	3.15
	5.0	4.87	4.19	3.18
	7.5	5.02	4.27	3.23
	10.0	5.16	4.36	3.27
15	0	5.98	5.11	3.50
	2.5	6.34	5.29	3.55
	5.0	6.72	5.49	3.61
	7.5	7.10	5.69	3.68
	10.0	7.47	5.91	3.77
20	0	7.02	6.05	3.93
	2.5	7.65	6.36	3.98
	5.0	8.31	6.69	4.05
	7.5	9.01	7.05	4.15
	10.0	9.72	7.42	4.27
100	0	9.33	15.53	12.25
	2.5	11.73	16.58	12.21
	5.0	15.75	18.36	12.18
	7.5	23.06	21.63	12.19
	10.0	36.05	27.54	12.42

Table 3 illustrates this effect for property investment.   Column 3
repeats the duration estimates of Table 2.   Column 5 approximates to
the case where changes in the interest rate are matched by equal

changes in the rental growth rate. Column 4 illustrates an
intermediate case of association. The effect on duration is
substantial and, in the extreme case, gives rise to durations which
are very comparable to bond durations. In this respect, it should be
noted that column 5 durations would apply if initial yields were
completely insensitive to interest rate changes. (The initial yield
can be thought of as the approximate difference between the discount
rate and the rental growth rate.) In the US, the early UK
experience with property did not occur and it is only in the last two
years that institutions have made significant investment in anything
like the UK commitment. A recent estimate of the scale of the
potential demand for US commercial property made by Zisler and
Ross:1987 forecasts $100 billion increase in institutional property
investment. Clearly this involvement will have a major impact on the
market, the composition of pension funds' portfolios and the way in
which the property assets behave.

5   How Risky is Property as a Long Term Investment?

In the earlier sections of this chapter, I have tried to show that,
judged by the criterion of long term fund management, property is of
particular use because it provides long duration assets. In this
special sense it is risk reducing. However, it must be admitted that
the immunisation argument is not yet widely accepted and many
researchers in the finance literature tradition would still expect
risk to be measured in terms of either systematic risk or total
variability. I turn first to the measurement of systematic risk.
 There are two ways in which we can justify systematic risk as the
appropriate measure. On a theoretical front, the Capital Asset
Pricing Model (CAPM) is often invoked as a central plank. As a
theoretical model it is very attractive because it is simple to
understand, it is simple to use and it is consistent with the basic
economic concept of 'equilibrium'. Unfortunately it is as useful as
other well known economic concepts such as 'perfect competition',
'economic rent' and 'capital': wonderful as logical constructs but
lacking just that linkage with reality to allow them to be seriously
tested or applied. In terms of testing, I do not know of any test of
the CAPM that has not shown it to be invalid or at best untested.
One recent paper (Board:1986) which looked at the UK stock exchange
established clearly that the CAPM had nothing to offer in explanation
of efficient portfolios. An earlier paper of the UK exchange by
Levy:1983 showed also that the empirical basis of the CAPM was
pathetically weak.
 However we do not need to rely on the CAPM in order to derive
systematic risk as an appropriate measure. The latest plaything of
the financial equilibrialists is a more sophisticated version dubbed
the Arbitrage Pricing Model (APT). This is a statistically-driven
theory as opposed to the economically static CAPM but comes with a
similar prediction that returns of a security (asset) will be
explained by its systematic relationship to one to five factors. The
connection is that the two theories are consistent if tests reveal
only one factor satisfactorily explains returns.

Unfortunately, empirical support for the APT is just as weak as it was for the CAPM. Coggin and Hunter:1987 reviewed two studies in the UK and concluded (p.37) that the '... support for the global null is striking', (For the benefit of those unaccustomed to research jargon, this phrase means that the authors found that the results simply destroyed the theory). In a study of the UK market Diacogiannis:1986 found that one could predict the number of factors that were consistent with the APT by looking at the size of the portfolios that were used to test the model. Small portfolios produced a one to two factor APT, larger portfolios produced more factors. Again this evidence is not supportive of the theory that one factor exists and does not justify in any way the argument that we should measure risk by the relationship between an asset and one factor (whatever it is).

However, we could jettison any claim to theoretical underpinning to systematic risk and merely argue that returns can be represented by an equation of the form

$$R(t) = A + B \times I(t) + e(t)$$

where A and B are constants
    I(.) represents an appropriate index
    e(.) is an error term for which
                $E(e) = 0$
                $E(e(t),e(t-k))=0$ for all non-zero k
                and $E(e(t),e(t)) = V$

As an empirical statement, this can be used if the estimated A and B are reasonably stable from one period to another, either in individual or in portfolio terms. Having established such a relationship, we cannot necessarily argue that B represents risk for investors unless we find that investors (a) hold portfolios of sufficient size that the variance of the residual error terms disappears and (b) that investors can predict the values of B from one period to another. If the second of these conditions does not hold, we would have the absurd position that investors could not predict how 'risky' the asset was that they contemplated buying. If the first position does not hold, the prices (and therefore the returns) in the market will reflect the effect of the residual error as well as the sytematic 'risk'. In other words, assets with the same value of B will not have the same returns. This possibility seems very plausible and is proposed by a number of writers including Levy:1980 and Logue and Merville:1973. Again we must withhold judgement until more research is published.

If the support for the systematic risk concept is weak in the studies of the relatively efficient financial markets, the support for the model in the relatively thinly-traded property market must rest on even less stable ground.

The alternative risk measure is total variability (or the standard deviation of returns). This always tends to appear in any analysis based on either small portfolios or segmented markets. It does not, however, illuminate the pricing of property assets greatly because the measured variability of returns from the direct property market appears too low relative to other assets. (See for example Ross and Zisler:1987, or Blundell and Ward:1987.) The returns used in most studies to date have relied heavily on valuations, not market transactions, and therefore are extremely difficult to turn into statements of return behaviour and dangerous to use as the basis for constructing models of risk/return trade-off. In both of the above studies the authors transformed the raw returns in order to correct the variability. The transformations were similar in both the method and their magnitude producing a significant increase in the estimated variability. There still remains a major problem because it would appear that investors were not investing in a way that was consistent with the return/variability trade-off of property when compared with other assets. My interpretation of the 'anomaly' is that the market is too thin to estimate reliable variabilities and I would predict that as more tradeable property assets appear, so the volatility of property returns will sharply increase. The support for this view can be found in the variability of shares in property companies listed on the stock exchange.

6   Conclusion

The aim of this chapter is to sensitise the practitioners and researchers in property to relevant issues which are still the subject of research in main stream finance. As I have tried to show, research into the usefulness of bonds and equities for long term funds has a direct and important relevance for property specialists, although considerable work remains to be done before the research can be commercially exploited. Research into risk in financial markets is also of immediate importance to the evaluation of property. What I hope is obvious from my discussion is that there are considerable opportunities to exploit in finding clearer answers to the questions raised. If property specialists and researchers do not invest time and other resources into researching the issues, the financial specialists will dominate the publication of significant research and more importantly the market place.

Appendix 1

Duration of property investment

1. Valuation of leasehold of mn years, reviewed at n years, rental growth rate = g, discount rate = r, initial rent = R_o.

$$V = \frac{R_o}{r} \left[\frac{(1+r)^n - 1}{(1+r)^n - (1+g)^n} \right] \cdot \left[\frac{(1+r)^{mn} - (1+g)^{mn}}{(1+r)^{mn}} \right]$$

2. Duration of leasehold

$$\frac{1 + r}{r} + \frac{n(1+r)^n}{(1+r)^n - (1+g)^n} + mn - \frac{n(1+r)^n}{(1+r)^n - 1} - \frac{mn(1+r)^{mn}}{(1+r)^{mn} - (1+g)^{mn}}$$

3. Duration of freehold

$$\frac{(1+r)}{r} + \frac{n(1+r)^n}{(1+r)^n - (1+g)^n} - \frac{n(1+r)^n}{(1+r)^n - 1}$$

4. Duration of leasehold with $dg/dr \neq 0$

$$\frac{(1+r)}{r} - \frac{n(1+r)^n}{(1+r)^n - 1} + \frac{(1+r)}{(1+r)^n - (1+g)^n} \left[n(1+r)^{n-1} - n(1+g)^{n-1} \, dg/dr \right]$$

$$+ mn - mn(1+r) \left[\frac{(1+r)^{mn-1} - (1+g)^{mn-1} \, dg/dr}{(1+r)^{mn} - (1+g)^{mn}} \right]$$

5. Duration of freehold with $dg/dr \neq 0$

$$\frac{(1+r)}{r} - \frac{n(1+r)^n}{(1+r)^n - 1} + \frac{n(1+r)}{(1+r)^n - (1+g)^n} \left[(1+r)^{n-1} - (1+g)^{n-1} \, dg/dr \right]$$

References

Ambachtsheer K P:1987, 'Pension fund asset allocation: in defence of a 60/40 equity/debt asset mix', Financial Analysts Journal, Sept–Oct pp.14–24.

Blundell G F and C W R Ward:1987, 'Property portfolio allocations: a multi-factor model', Land Development Studies, 4, pp.145–156.

Board J L G:1986, 'A Test of the Efficiency of various Market Indices', Working Paper, University of Reading.

Boulding K E:1936, 'Time and Investment', Economica, May.

Brown G E:1984, 'Assessing an all-risks yield', Estates Gazette, Vol.269, pp.700–706.

Chua J H:1984, 'A closed-form formula for calculating bond duration', Financial Analysts Journal, May–June, pp.76–77.

Coggin T D and J E Hunter:1987, 'A meta-analysis of pricing "risk" factors in APT', Journal of Portfolio Management, Fall 1987, pp.35–38.

Diacogiannis G P:1986, 'Arbitrage Pricing Model: A critical Examination of its Empirical Applicability for the London Stock Exchange', Journal of Business Finance and Accounting, Vol.13 No.4, pp.489–504.

Fischer D E and R J Jordan:1987, Security analysis and portfolio management, (Prentice-Hall, 4th Edition).

Fraser W:1986, 'The target return on UK property investments', Journal of Valuation, 4, pp.119–129.

Leibowitz M L:1986, 'Total portfolio duration: a new perspective on asset allocation', Financial Analysts Journal, September–October.

Levy H:1980, 'The CAPM and beta in an imperfect market', The Journal of Portfolio Management, Winter, pp.5–11.

Levy H:1983, 'The capital asset pricing model: theory and empiricism', Economic Journal, March, pp.145–165.

Logue D E and L J Merville:1973, 'General model of imperfect capital markets', Southern Economic Journal, Vol.40, pp.224–233.

Redington F M:1952, 'Review of the principles of Life-Office Valuations', Journal of the Institute of Actuaries.

Ross S A and R C Zisler:1987, 'Managing real estate portfolios: a closer look at equity real estate risk', Goldman Sachs Working Paper.

Zisler R C and S A Ross:1987, 'Stock and bond market volatility and real estate's allocation', Goldman Sachs Working Paper.

CHAPTER 9

ADVANCES IN PROPERTY INVESTMENT THEORY

1   Introduction

Each of the previous chapters in this volume has been dealing with a specific topic within the theme of property investment analysis and the application of techiques of risk analysis within the main stream of financial theory to property investments.

The purpose of this chapter is to draw together some of the issues which were identified at the seminar where the authors delivered papers in these subjects. Discussion of the papers by delegates was active and informed. This led to clarification of many of the issues: in some cases to a concensus, in other cases to a continuing disagreement, but based on common concepts or factual evidence.

Given the nature of the subject an attempt to synthesise the opinions of all of the participants on all of those issues would clearly be difficult if not impossible. The facets of debate which are further analysed here have therefore been selective. It is hoped that they will be considered to be the most important, whether or not the argument is adequately reflected. As a result the pointers to future areas of research will in turn tend to be selective and narrower than the real range of information needs. However, it is again believed that the signposts will be to those areas where the most useful and practical research can be undertaken.

In approaching the synthesis of the material contributed by the various authors and the informed comment on the material each topic is examined in turn then those areas upon which consensus appears to have been achieved are affirmed. Those issues which run through subject areas are re-appraised and then suggestions as to future work are made.

2   The Issues

A considerable amount of discussion over the past decade has focused on the topics of market valuations and investment appraisals of property investments. It is very common to find articles in property journals advocating the use of explicit discounted cash flow models as improved tools of analysis for investment appraisals. However, these developments, though undeniably useful, have tended to concentrate on manipulations of deterministic models to accommodate new variables such as depreciation rates and the terminal value of the property at the end of an assumed holding period, but very little research has been done in quantifying the inputs into the models. This situation is understandable, given the paucity of data available in the public domain. We have made some progress in analysing the impact of depreciation on commercial property. It is only now we are beginning to recognise that portfolio considerations are important in property investment analysis and that investment appraisal must include the twin measures of risk and return.

It is generally recognised that the analysis of property investments should be more explicit and such analysis should include a formal treatment of risk. However, there is much concern over how this should be done given the nature of the asset and the market.

It is hoped that the ensuing discussion of the major issues arising out of each of the chapters in this book may be useful in focusing attention on the nature of the problems and on the importance of finding satisfactory solutions.

3 Valuation and Investment Appraisal

Chapters 2 and 3 of this book highlight the ongoing debate concerning the choice of technique ('conventional' method versus 'DCF') for market valuations. While there is agreement that investment appraisals must be based on explicit DCF analysis, the issue of market valuations remains unsettled. The indications are that the DCF approach is acceptable for market valuations of certain types of property investments where comparable evidence is not likely to be freely available. Non-standard freehold reversions and short leaseholds are typical examples of such investments.

The valuation of investments for which comparable evidence is available in sufficient quantity is relatively straight forward and both conventional and DCF methods can be employed in such situations. Indeed, the two approaches if properly used are not likely to produce significantly different values. The problem, however, remains with the valuation of investments where such evidence is not available. The current DCF models are based on simplistic assumptions and unless they conform to the perceptions of investors in the market, the resulting valuations may not be a true reflection of market values. This point also raises the question, 'Does the valuer adjust information in the way the market adjusts it'?

A major issue that arises from the topics of market valuation and investment appraisal concerns the appropriate discount rate or the target rate for property at the market, sector and asset level. This point raises two questions.

(a) What rate of return should investors expect from an investment?
(b) What rate of return do investors expect?

The first question is important from the point of view of investment appraisal. The answer to this question might be found in developments in Capital Market Theory. The risk adjusted expected rate of return can be derived from the Capital Asset Pricing Model (CAPM). Indeed, Brown in Chapter 7 argues that the proper risk adjusted discount rate could be derived from the CAPM based on an estimate of the market (or systematic) risk of the investment. In a recent paper in the Journal of Valuation, Brown (Brown:1987) describes a model for the estimation of the market risk of individual properties and portfolios. These are useful developments but there remain a number of theoretical and empirical problems which are yet to be resolved. So further research is required in the area of investment appraisal rules under conditions of risk before satisfactory solutions can be found.

It .s more difficult to find a satisfactory answer to the second question. The target rate is normally deemed to be 2% above the redemption yield on long dated government bonds for good quality investment property. But a value for the risk premium is not settled. It is argued that the appropriate risk premium must tend to vary between different properties and over time. There are arguments in favour of a nil or negative risk premium (see Fraser: 1986) as well as support for a 2% risk premium (See Brown:1985).

A related issue is the question as to whether dated gilts constitute a risk free investment. In nominal terms gilts can be considered as risk free if held to redemption. However, due to interest rate movements, the periodic returns fluctuate and from this view point gilts must be regarded as risky. Without a totally acceptable definition of risk, it is difficult to decide what constitutes a risk free investment. However, the redemption yield on a gilt may be used as a surrogate measure for the risk free rate over the period to maturity.

Given these diverse views on the target rate of return, a survey of institutional investors' criteria for investment in property could be a very useful starting point in finding answers to these questions.

4   Depreciation

In recent years, investors have been greatly concerned with the impact of depreciation on commercial property. There is concern that current yields on property investments may not adequately reflect the impact of depreciation on rental growth and consequently on future performance.

Recently, several research projects have been undertaken to investigate the causes and the impact of depreciation. In Chapter 4, Baum identifies the major components of depreciation as internal planning arrangements, specification and external appearance. Each was found to be more significant than physical deterioration. These findings are useful in the design and construction of new developments where the emphasis should be on the provision of buildings with maximum flexibility so as to minimise the impact of depreciation.

The estimation of depreciation rates are very important for investment appraisals. However, the problems of estimation are exacerbated by the imbalance between demand for and supply of commercial space which tends to mask the impact of depreciation. For example, at times of extreme space shortage, it is possible to let buildings at high rents. It is probable that depreciation rates may exhibit cyclical patterns. Is it therefore, realistically possible to make use of these trends in the future in the estimation process?

Different buildings may depreciate at rates that are widely different. For example, a Georgian Mayfair property can be more resilient against depreciation than a 1960's office building. This would mean that average rates of depreciation cannot be used in the appraisal of different properties even if they are situated in similar locations. It is, therefore, inevitable that some element of subjectivity will be associated with the estimation of depreciation rates.

The impact of depreciation may now be regarded as information that is publicly available. Property yields have increased slightly in the recent past but it is difficult to ascertain the extent to which this upward movement in yields can be attributed to an increasing awareness among investors of the impact of depreciation. In Chapter 7, Brown argues that the property market operates in a broadly efficient manner and therefore, the impact of depreciation should be reflected in values. However, Salway's (Salway:1986) study on depreciation indicates that for offices in certain locations this does not appear to be the case.

A further point that merits consideration is whether there should be tax allowances on buildings for depreciation, and the likely effects of such allowances on investments in property.

5   Forecasting

Forecasting is an area of enormous importance to property investment analysis. But it is an area that has received very little attention from property researchers. The paucity of research in this area can be attributed to the fact that the traditional expertise of the surveyor does not include any knowledge of econometric modelling which is essential for the development of forecasting models. Consequently, the property profession must rely heavily on the expertise of analysts trained in the disciplines of econometrics, mathematics and statistics.   A further problem with the forecasting of property market variables is the lack of data both in terms of quality and quantity. The position is improving with major firms of surveyors developing large data bases, but it will take some years before the data becomes available in sufficient quantity to enable useful empirical work to be carried out.

The forecasting models presented in Chapter 6 of this book show how rental values and yields could be forecast in the short term. One of the main problems with forecasting is the identification of adequate explanatory variables.   The degree of explanatory power of some of the variables will change over time necessitating frequent revisions to the models.   For example, until 1979, property yields and redemption yields on long dated gilts were highly correlated but the linkage between these two variables appears to have changed in the 1980's.

The models forecast rental values and yields over a period of one year.   These forecasts can be useful in the timing of decisions concerning acquisitions/disposals of property investments.   However, more research is required in the forecasting of long term trends in variables such as rental growth.   This would involve the assessment of the impact of technological change, demand and location, on expected level of growth.

6   Risk Analysis

Although it is generally recognised that there should be a move towards formal risk analysis in the appraisal of property investments, there is little agreement on how this should be done.

This lack of concensus is understandable given the widely different methods that have been suggested for the evaluation of risk. For example, should risk be measured by the variability of returns (standard deviation of the internal rate of return) based on subjective forecasts of a property's future cash flows or should it be measured objectively based on concepts of systematic risk? The standard deviation of historic rates of return is suggested as yet another measure of risk. Alternatively, should we focus our attention on the concern of major investors in the market, that of matching assets with liabilities?

A related but different issue concerns the frequency of performance measurement. Given the relatively longer holding periods of property investments and the difficulties with estimating values over short intervals of time, is it useful or realistic to produce measures of returns over such short intervals as monthly or quarterly periods? These are some of the issues that will increasingly engage the attention of both investors and analysts in the near future.

Dealing with risk at the individual asset level, Analytic or Simulation Models (i.e Hillier or Hertz models) can be employed depending on the complexities of the situation. The output from such an analysis, the expected value and the variance of the NPV or IRR provides more information for the investor than the single valued estimate produced by deterministic models.

Probabilistic models are relatively easy to use if we make simplifying assumptions that the variables are normally distributed and that there are no dependencies between them. However, the aim of risk analysis is to be explicit, but this may conflict with the objective of adopting simplified techniques to gain acceptance in practice if this means that the assumptions also have to be simplified. The models can be refined to cope with dependencies but the main difficulty lies in assessing the degree of dependence. This is an empirical problem and analysis of historical data could be of some value. For example, an analysis of historical rental growth rates can be useful in assessing the degree of serial correlation that exists and research could be directed into this area. It is also important to recognise that ignoring the dependencies will cause the variance of the NPV or the IRR to be understated.

7   Portfolio Theory and Property Investment Analysis

Gerald Brown's research is highly significant because it is one of very few studies undertaken in this country in the area of empirical analysis of risk and return in the commercial property market. In Chapter 7, he demonstrates the application of portfolio theory and the capital asset pricing model to property investment analysis. His observations that traditional investment advice is likely to lead to economically correct decisions only by chance and that the present systems of property performance say very little about real performance and are of doubtful value in guiding investment strategy, clearly suggest the need for more fundamental research into the property market.

The above observations raise a number of issues that need to be addressed in tackling the problems that are involved in property investment analysis. Portfolio theory and the concept of diversification are applicable to all assets including property, provided the necessary inputs can be correctly estimated. However, it is fair to say that the majority of the investors and practitioners in the property world have limited knowledge of mathematics and this is one of the reasons for the general reluctance on their part to accept models that require a reasonable knowledge of mathematics to understand and implement. For example, they recognise the benefits of diversification but not in terms of efficient diversification as defined by Markowitz.

Some of the empirical work done by Brown relies on the Capital Asset Pricing Model. The CAPM is derived from a set of stringent assumptions and it is debateable whether such assumptions can be extended to the property market. It may be argued that the final test of a model is not how reasonable the assumptions behind it appear but how well the model describes reality. The CAPM has been subject to extensive empirical testing but the evidence does not provide strong support for it (Peasnell: 1986). However, the CAPM has not been subjected to an unambiguous empirical test (Roll:1977). In a critique of the asset pricing theory's tests, Roll concludes, 'two parameter asset pricing theory is testable in principle; but arguments are given here that (a) no correct and unambiguous test of the theory has appeared in the literature and (b) there is practically no possibility that such a test can be accomplished in the future.'

Ward (see Chapter 8) states that the empirical support for the model in the UK Stock Exchange was disappointingly weak. He also reports that empirical support for the Arbitrage Pricing Model (APT) is just as weak as it was for the CAPM.

A further difficulty with property analysis is that the returns are valuation derived. Brown (Brown:1985) argues that valuations are a good proxy for prices in the institutional investment market and based on this assumption the property market is efficient at the weak form level. These findings are not only interesting but crucial and must be followed up with further research.

The contents of Chapters 7 and 8 or this book have raised a number of important issues concerning the risk of property which may be encapsulated in the following questions:

(a) Are valuations a good proxy for market prices?
(b) How efficient is the institutional property investment market in the UK?
(c) How useful is the Capital Asset Pricing Model in determining equilibrium values of property investments?

The application of Portfolio Theory and Asset Pricing Models to property investment analysis is a very recent development and further research is required before firm conclusions can be made on their usefulness for the analysis of property.

8 Property as a long term investment – the contribution of duration

Unlike the bond market, the concept of 'duration' is not very familiar to many in the property market. It is generally recognised that one of the major concerns of institutional investors is the problem of matching. Ward (see Chapter 8) demonstrates that property assets tend to have long durations and should be of importance to long term funds since their liabilities have long durations.   If the immunisation argument gains wider acceptance then decisions concerning the composition of pension funds portfolios may be influenced by the 'duration' of the assets. Furthermore, if satisfactory solutions to some of the problems associated with the use of the systematic risk measure are not easily found there could be a move in this direction for the analysis of portfolios of long term funds. However, as Ward points out a considerable amount of both theoretical and empirical research has to be done in this area before it can be commercially exploited. Property researchers should therefore make a start in this direction by researching into the duration of property assets. For example, the correlation between interest rates and expected rental growth rates may be investigated and this could enable us to study the sensitivity of capital values to interest rate movements.

9 Research – some pointers

In contrast to financial investments there is paucity of research in property investments. The insitutional property market can no longer be viewed in isolation from the capital markets. With increasing need to justify investment decisions, the pressure for analysing property using techniques that have been developed in the financial markets is likely to increase. This would mean that investment surveyors must acquire a sound knowledge of modern finance and where appropriate be able to apply modern techniques, developed in the theory of finance, to property investment analysis. If there are reasons to believe that certain methods are inappropriate then these must be forcefully and convincingly presented and more importantly attempts must be made to develop suitable alternative methods that are conceptually sound and capable of practical application.

   In the context of the issues that have been highlighted in this chapter, it is evident that a great deal of the material presented in each of the chapters is worthy of further research. However, the following are some of the aspects of property investment analysis which we believe to be important for further research:

(a) Investment criteria used by institutional investors
There is a need to examine the criteria used by pension funds and insurance companies in the selection of assets. This study might give valuable insights into the risk preferences of investors which are useful in developing models for investment analysis. For example, we might be able to establish whether the concept of immunisation is important in asset selection and pricing.

(b)   Property market efficiency
The efficiency of the property market is generally regarded as the lowest amongst all other markets. Gerald Brown however, argues that it is efficient at the weak form level based on the assumption that valuations are a good proxy for prices.

Therefore we may attempt to find answers to the following questions:

1.   Are valuations a good proxy for prices?
2.   Are there sufficient arbitrage opportunities for abnormal returns to be earned on a consistent basis and can such situations be easily identified?

(c) The importance of property in investment portfolios
It has been argued that property investments have relatively low covariance with the market and therefore property is a good medium for diversification. From the view point of matching, it is suggested that property provides an efficient source of long duration matching.   Further research into these areas will be useful in order to examine the role of property in a portfolio. The potential for risk reduction by sector, location and age should also be investigated.

(d)   Construction of a suitable property index
The usefulness of performance measurement is limited by the absence of reliable indices. Research may be undertaken towards the construction of a reliable property index which accurately tracks movements in the market and is free of the problems of smoothing and serial correlation.

(e) Forecasting
Research into improved methods of forecasting market and portfolio variables.
     Research into the identification of plausible economic indicators and estimating the relationship between property and the statisticially related variables will be useful. It is also essential to consider over what longer period forecasting can be made, as the present forecsts on rental growth or yield do not extend beyond an year. There are difficulties but more research should be directed to this problem.

10 Data Collection and Dissemination

Property research requires a better data base. With the development of performance measurement the situation is improving. Research into some of the topics discussed in this chapter requires extensive time series data of property returns.   The major property firms in London have developed substantial data bases and have already started researching into some aspects of property. However, it is important that researchers from outside, particularly, those from the academic institituions be allowed access to data. Their involvement in empirical research can greatly facilitate research into property and the benefits to the property industry from such an involvement could

153

be very significant.

11 Conclusion

The purpose of this chapter has been to identify key issues and areas
of research importance and to stimulate discussion.   Whilst there
was common ground on many of the issues at the seminar there were
also a number of unresolved issues a fact that this chapter has
attempted to reflect.

References

Brown G:1987 'A certainty equivalent expectations model for estimating the systematic risk of property investment', Journal of Valuation, Vol.6, p.17

Brown G:1985, 'Property Investment and Performance Measurement', Journal of Valuation, Vol.4 p.33

Fraser W:1986, 'The target return of UK property investments', Journal of Valuation, Vol.4, p.119

Peasnell K:1986, 'The Capital Asset Pricing Model', in Firth M and Keane S M (eds), Issues in Finance, Phillip Allan, p.28

Roll R:1977, 'A Critique of the Asset Pricing Theory's Test, Part 1: on past and potential testability of the theory', Journal of Financial Economics, March, p.129

Salway F:1986, Depreciation of Commercial Property. CALUS research report

ADVANCES IN PROPERTY INVESTMENT THEORY

TWO-DAY SEMINAR, 18 AND 19 FEBRUARY 1988

Delegate List

Andrews, Roy	Deputy Surveyor, Pearl Assurance plc, London WC1
Arnison, C J	City of Birmingham Polytechnic, Birmingham
Bell, Robin	Accountant and Receiver General, The Crown Estate Commissioners, London SW1
Blundell, Gerald	Partner, Jones Lang Wootton, London W1
Borley, Geoff	Property Portfolio Manager, Shell Pensions Trust Ltd, London SE1
Bornand, David	Senior Lecturer, Dept. of Urban and Regional Studies, Sheffield City Polytechnic, Sheffield
Carr, David	Financial Analyst, Hillier Parker Financial Services, London W1
Chegwidden, Neil	Research Assistant, Hillier Parker Research Ltd, London W1
Clifton Brown, James	Hillier Parker, London W1
Conellan, Owen	Course Leader, Estate Management, Kingston Polytechnic
Crosby, Neil	Lecturer in Valuation, Department of Land Management, University of Reading, Reading
Cunis, Nigel	Partner, Gerald Eve, London W1
Dent, P	City of Birmingham Polytechnic, Birmingham
Dixon, Timothy	Lecturer, College of Estate Management, Reading
Edwards, Christopher	Prudential Portfolio Managers Ltd, London EC1
Flint, Alan	Deputy Fund Manager (Property), London Residuary Body, London SE1
Forster, Miss Sue	Hillier Parker, London W1
French, Nick	Valuation Lecturer, City University, London EC1
Gilchrist, Robert	Surveyor, MIM Ltd, London EC2
Gillespie, David	Director of Information Technology, Cluttons, London SW1
Grimble, Michael	Chief Economist, Norwich Union Insurance Group, Norwich
Hargitay, Stephen	Principal Lecturer, Department of Surveying, Bristol Polytechnic, Bristol

Haslam, Susan	Senior Lecturer, Department of Surveying, Liverpool Polytechnic, Liverpool
Hoare, Geoff	Hillier Parker, London W1
Hynes, Christopher	Crown Estate Surveyor, The Crown Estate, London W1
King, Robert	The Hammerson Group, London W1
Knight, Ray	Director, Town and City Properties Ltd
McCluskey, William	Lecturer, University of Ulster at Jordanstown, N Ireland
McIntosh, Angus	Head of Research, Healey and Baker, London W1
McKinnell, Keith,,	Department of Surveying, University of Hong Kong
Mallinson, Michael	Chief Surveyor and Property Director, Prudential Portfolio Managers Ltd, London EC1
Morrell, Guy	Property Research Analyst, Prudential Portfolio Managers Ltd, London EC1
Shearmur, P G	Crown Estate Surveyors Branch, The Crown Estate, London SW1
Sieracki, Miss Karen	Property Investment Assistant, Electricity Supply Pension Scheme, London SW1
Smith, Julian	Principal Assistant Senior Surveyor, Electricity Supply Pension Scheme, London SW1
Speakman, John	Senior Property Surveyor, Legal and General Property Ltd, London NW1
Strang, Andrew	Investment Surveyor, Hill Samuel Property Services Ltd, London EC2

INDEX

Negative yield gap 4
Net present value 64, 83, 90, 92, 110, 114, 122, 131
Nominal returns 139
    "   value 136

Obsolescence (see also Depreciation) 6, 14, 49, 59, 72, 116
Open market value 112

Pension funds 2, 5, 6, 135, 138, 141
Performance 121, 130
    "   measurement 67, 110, 122
Portfolio
    "   investment 2, 5, 9, 93, 123, 153
    "   structure 5, 98
    "   theory 110, 112, 150
Probability 78, 85, 88-94
    "   analysis 70-75, 150
Property
    "   boom 2, 9, 113
    "   companies 3, 4, 5, 108, 143
    "   market 4, 10, 15, 49, 63, 74, 97, 100, 113, 116, 129, 142,
               153
    "   sector 28, 130, 131, 132, 147
Price/Earnings ratios 13

Redemption yield (see Yield)
Refurbishment 57-60, 72, 133
Regression analysis 59, 102, 103
Rent
    "   growth 9, 14-17, 32
    "   industrial 99, 100
    "   office 49, 56, 98
    "   rack 9
    "   retail 98-104
Return
    "   abnormal 119, 122, 123, 128, 132
    "   expected 117, 120
    "   market 124, 129
    "   rate of 6, 14, 28, 97, 120, 128, 147
    "   target 14, 17, 24, 63, 123
Reversions 18, 19, 20, 25, 29, 31, 74, 75, 84
Risk 10, 15, 18, 28, 31, 67, 70, 71, 93, 94, 111, 112, 116, 122, 124,
      134, 136, 138, 141, 146, 150
Risk adjusted discount rate 24, 29, 79, 82, 115, 122, 147
    "   analysis 11, 37, 70, 71, 74, 75, 88, 94, 149
    "   aversion 81, 114
    "   averters, seekers 114
    "   class 28, 115, 122, 130
    "   free rate 15, 61, 79, 80, 112, 114, 120
    "   non-market risk 112
    "   premium 15, 24, 28, 48, 61, 79, 120, 148
    "   systematic (market) 112, 115, 119-23, 141, 142, 147
    "   unsystematic (specific) 112, 119, 120, 125
Riskless asset 114, 129, 148

Royal Institution of Chartered Surveyors (RICS)   10, 117

Sale and leaseback. 4
Scenarios   77-79, 138
Security market line   113, 114, 115
Sensitivity analysis   31, 40, 71, 74-79, 82, 140, 152
Serial correlation   121
Sharpe Model (see Model)   111
Short leasehold   20, 37, 74, 139, 147
Sliced income approach   82
Standard deviation   78, 80-91, 111, 120, 150
Stock Exchange   141, 151
Stock market   6, 9, 72, 98, 142
Superannuation funds   110, 152

Trend lines   103, 108

Unitised property   114, 117, 143
Upside potential   121, 129, 131
Utility   114, 115

Valuation
   "   market   13, 15, 17, 18, 24, 146, 147
   "   method of   6, 13, 14, 21, 23
   "   tables   10
Variables
   "   dummy   102, 103
   "   market   15, 98
   "   portfolio   98 ·

Years purchase   15-21, 24, 26
Yield
   "   'all risks'   16, 24, 36, 48, 70, 97, 115
   "   premium (or discount)   17, 18, 72
   "   redemption   4, 24, 28, 29, 48, 137, 138, 148
   "   reversionary   18

Symbolic Dictionary

1. C – Risk free inflation prone opportunity cost rate or nominal risk free rate of return.

2'. d – Periodic (compound) rate by which rental growth is eroded by depreciation.

3. D – Duration of an investment

4. $E(r_p)$ – Expected return on the portfolio.

5. g – Periodic growth in rental values net of depreciation

6. g_m – Periodic growth in rental values of brand new buildings

7. i – Reward for anticipated inflation

8. l – Reward for losing capital or real risk free rate of return

9. n – Rent review period

10. p – risk premium

11. r – Opportunity cost of capital or equated yield or internal rate of return

12. r_f – risk free rate of return

13. r_m – return on the market portfolio

14. R – full rental value

15. V – capital value of an investment

16. Y – All risks yield or income return

17. B_p – Market risk of systematic risk of the portfolio

For Product Safety Concerns and Information please contact our EU
representative GPSR@taylorandfrancis.com
Taylor & Francis Verlag GmbH, Kaufingerstraße 24, 80331 München, Germany

www.ingramcontent.com/pod-product-compliance
Ingram Content Group UK Ltd.
Pitfield, Milton Keynes, MK11 3LW, UK
UKHW021610240425
457818UK00018B/484